COPPER IN ANIMAL WASTES
AND SEWAGE SLUDGE

Commission of the European Communities

COPPER IN ANIMAL WASTES AND SEWAGE SLUDGE

Proceedings of the EEC Workshop organised by the Institut National de la Recherche Agronomique (INRA), Station d'Agronomie, Bordeaux, France, and held at Bordeaux, October 8–10, 1980

Sponsored by the Commission of the European Communities, Directorate-General for Agriculture and Directorate-General for Research, Science and Education

Edited by

P. L'HERMITE and J. DEHANDTSCHUTTER
Commission of the European Communities, Brussels

D. REIDEL PUBLISHING COMPANY

DORDRECHT : HOLLAND / BOSTON : U.S.A.
LONDON : ENGLAND

Library of Congress Cataloging in Publication Data
Main entry under title:

Copper in animal wastes and sewage sludge.

At head of title: Commission of the European Communities.
1. Swine—Feeding and feeds—Congresses. 2. Copper in animal
nutrition—Congresses. 3. Swine—Manure—Environmental aspects—Congresses.
4. Soils—Copper content—Congresses. 5. Copper—Toxicology—Congresses.
6. Plants, effect of copper on—Congresses. 7. Soil pollution—Congresses.
8. Sewage sludge—Environmental aspects—Congresses. I. L'Hermite, P.
(Pierre), 1936–   . II. Dehandtschutter, J. (Jacques). III. European
economic community. IV. Institut National de la Recherche Agronomique
(France). V. Commission of the European Communities.
SF396.5.C66   636.4'0852   81-7338
ISBN-13:978-94-009-8505-6   e-ISBN-13:978-94-009-8503-2
DOI: 10.1007/978-94-009-8503-2

Publication arranged by
Commission of the European Communities,
Directorate-General Information Market and Innovation, Luxembourg

EUR 7196
Copyright © 1981 ECSC, EEC, EAEC, Brussels and Luxembourg, 1981
Softcover reprint of the hardcover 1st edition 1981

Proceedings prepared by
Janssen Services, 14 The Quay, Lower Thames Street, London EC3R 6BU

LEGAL NOTICE

Neither the Commission of the European Communities nor any person acting on behalf of the
Commission is responsible for the use which might be made of the following information.

Published by D. Reidel Publishing Company
P.O. Box 17, 3300 AA Dordrecht, Holland

Sold and distributed in the U.S.A. and Canada
by Kluwer Boston Inc.,
190 Old Derby Street, Hingham, MA 02043, U.S.A.

In all other countries, sold and distributed
by Kluwer Academic Publishers Group,
P.O. Box 322, 3300 AH Dordrecht, Holland

D. Reidel Publishing Company is a member of the Kluwer Group

CONTENTS

PREFACE

This publication constitutes the Proceedings of a Workshop held in
Bordeaux (France) on 8th - 10th October, 1980, under the auspices of the
Commission of the European Communities, as part of the CEC research
programme on 'Effluents from Livestock' and the Concerted Action COST
68 bis 'Treatment and Use of Sewage Sludge'.

Major changes have taken place in livestock production techniques in
recent years. One of the most important developments has been in the field
of animal nutrition. Animals are fed to gain maximum liveweight in close
relationship with market requirements for carcass and meat quality. With
regard to pig production, dietary formulation is based on scientific
principles and feed includes a large variety of ingredients to supply
optimum feed rations for 'standardised' animals.

In order to increase growth rate and to improve feed conversion,
copper is added to the rations of fattening pigs in a number of countries,
in accordance with the provisions of Council Directives concerning additives
in feedingstuffs, as last amended by the 23rd Commission Directive of
4th July 1980.

In accordance with these considerations, the subject of this Workshop
covered the following items:

a) Animal nutrition   : to consider how much copper is needed in pig
                          feeding;

b) Animal physiology   : to consider the influence of copper on pig growth
                          and where the animal stores it;

c) Agronomy   : to evaluate the risks to the soil-plant-animal
                          system in introducing copper in animal feed rations;

d) Effects on consumers: to assemble information on the possible effect on
                          consumers of eating meat from pigs receiving extra
                          dietary copper.

The Commission of the European Communities wishes to thank:

- the local organiser, Dr. C. Juste and his co-workers for the
excellent organisation of this Workshop;

- the scientific chairman, Dr. J.K.R. Gasser, who provided his
scientific expertise for the success of the meeting;

- the Chairmen of the sessions;

- the experts of Working Parties 2 (Chemical Pollution of Sludge) and
5 (Environmental Effects of Sludge) of the concerted action on
'Treatment and Use of Sewage Sludge', and their co-ordinators,
Dr. R. Leschber and Dr. G. Hucker respectively;

- all the participants who contributed fruitfully to this Workshop
through their scientific knowledge.

P. L'Hermite   J. Dehandtschutter

OPENING REMARKS

J.K.R. Gasser *(UK)*

Mr. G. Wansink, who is the Chairman of the Farm Wastes Committee, would normally be opening the Workshop but unfortunately he cannot be with us and therefore he has asked me as Scientific Chairman to undertake this duty.

On behalf of the Farm Wastes Committee I welcome you all here, and since this is a joint meeting, I also welcome you on behalf of the Treatment and Use of Sewage Sludge Committee of DG XII. Some of you will know that the activities of the Farm Wastes Committee under the Standing Committee for Agricultural Research (SCAR) will end formally at the end of this year because the funding will cease. This is, therefore, the last meeting organised by the Farm Wastes Committee and I think it is particularly appropriate that we should be discussing a joint problem. The venue was chosen with care as you may have realised because copper and Bordeaux, Bordeaux and vineyards, vineyard soils and copper are all linked.

It is now my privilege and pleasure to ask our host Institute to welcome you in the persons of M. Bulit, Administrateur, Centre de Recherches de Bordeaux and Dr. Delas, Director of the Institute.

WELCOME

J. Bulit *(Administrateur, Centre de Recherches de Bordeaux)*

It is with great pleasure, on behalf of the Scientific
Administration of the Institut National de la Recherche
Agronomique, that I welcome you to Bordeaux, and thank you for
coming to this Symposium.

It would perhaps be useful to give you a brief outline
of the INRA research centre at Bordeaux. I feel it is necessary to emphasise that the scientific organisation of INRA rests
essentially on specialised disciplines and is, therefore, a
vertical structure. In order to avoid isolation, with scientists lacking contact across the various disciplines, our
Institute is unusual in that it has, joined to the vertical
structure, a horizontal structure based on the Centre de
Recherche Régional. Each Centre de Recherche Régional brings
together various disciplines which, by being under one centre,
are able to collaborate. Such collaboration is based on
central topics and thus lends a certain homogeneity to a centre.

The Centre de Bordeaux concentrates on various main
topics: vines, fruit trees (and particularly, walnut trees),
forestry and maize. These principal areas of research concern
scientists in all the disciplines which form part of the Centre
de Bordeaux. These departments are: agronomy (the department
which is your host for the next three days); plant improvement,
part of which is devoted to vines and part to fruit trees;
plant pathology; zoology-entomology; molecular and cellular
biology; plant physiology; mushroom research; and a laboratory
for plant analysis.

For administrative purposes, the Centre also includes
other installations in the south west: the Continental Hydro-
biology Station near the Pyrenees; two forestry research
laboratories which are about 20 km away; (an ecology and forestry laboratory and a laboratory of forest-trees improvement); a

station for research into foie gras and geese. There are also experimental areas which cover some few hundred hectares.

That gives you some idea of the main preoccupations which bring together to research workers of the Bordeaux centre. I have spoken of vines, forests and fruit trees. I feel it would be good to round off by saying that research at the Centre de Bordeaux is accompanied by a certain gastronomy - perhaps as a result of regional influence - Bordeaux being very well placed in France, gastronomically speaking. The hydro-biology station devotes a lot of time to salmon and trout and the people at the goose research station are very interested in goose, duck and foie gras production. There are also edible mushrooms in Bordeaux, truffles and Bordeaux cèpes. Together with the wine this makes a whole range of attractions which I hope will make your stay here extremely pleasant.

You are all concerned with the problems posed by copper, and having read the programme, I see that it is above all the toxic effects which will excite your interest. As a patholo-gist I would ask you to remember copper in the guise of 'bouillie bordelaise' or Bordeaux mixture, for it was in Bordeaux, about 100 years ago, that the Bordeaux mixture was perfected by Millardet and Ulysse Gayon, and you will not be unaware that copper sulphate, Bordeaux mixture, is still a reliable fungicide, particularly for vines, and is still widely used. Despite recent chemical discoveries, the Bordeaux mixture is still efficacious and so copper can be beneficial, even if its accumulation in the soil over a century has led to certain difficulties.

Once again, I bid you welcome to Bordeaux and I hope you have three days which are full of hard work, fruitful, and above all, enjoyable.

J.K.R. Gasser (UK)

Thank you very much, M. Bulit for your warm welcome to

the Institute. Now, M. Delas will officially welcome us to
the Symposium.

J. Delas *(Directeur, Station d'Agronomie)*

    I am happy to welcome all the participants to the EEC
Workshop on problems encountered with copper. The Station
d'Agronomie, INRA de Bordeaux has been concerned with copper
toxicity in vineyard soils for over 20 years. This problem
seems to be solved now as far as viticulture is concerned but
copper accumulation occurs in certain soils because of the
addition of organic wastes which are rich in copper from
slurries and sludges. The positive results we have obtained
in the past are probably still relevant and this is why Bordeaux has been chosen for this workshop.

    I am convinced that this meeting, organised by my friend,
M. Juste, will be valuable to everyone and that some interesting studies will be presented. I only hope that the weather
improves sufficiently to allow the afternoon visit to the
experimental fields and that you will be able to enjoy discov-
ering one of the most prestigious vineyards in the world, the
Medoc.

    Thank you.

SESSION I

ANIMAL NUTRITION AND PHYSIOLOGY

Chairman: A. Aumaitre

TWENTY FIVE YEARS OF WIDESPREAD USE OF
COPPER AS AN ADDITIVE TO DIETS OF GROWING PIGS

R. Braude

C.A.B. Pig News and Information,
Lane End House, Shinfield, Reading, UK.

Twenty five years have passed since Barber, Braude, Mitchell and Cassidy (1955) reported that performance of growing pigs was improved by supplementing their diet with copper sulphate, supplying 250 mg Cu/kg diet. A great amount of research endeavour followed throughout the world and established beyond any reasonable doubt that an addition of 200 - 250 mg Cu/kg diet results in increasing the profitability of pig keeping, without having any known adverse effects on the pig or on the consumer of pig meat and meat products (see reviews by Braude, 1965 and 1975; Wallace, 1967; Meyer and Kroger, 1973; UKASTA, 1978; Omole, 1980). A very forthright support for the use of copper in pig feeding has recently been made by Wilson et al. (1979) in a paper on 'Copper as an inexpensive growth promoter for the pig'. There is very little that one can add to elaborate on the existing evidence, but one can stress the fact that during the last 25 years, in the United Kingdom, at least 200 million pigs received diets supplemented with copper, and, as far as I know, not a single person or animal has suffered from it in any way. It is much more difficult, virtually impossible except as a speculation, to estimate the usage throughout the world, but I would venture a suggestion that at least 2 billion pigs received diets supplemented with copper.

There have been a few reports in the world literature reporting some adverse effects due to copper supplementation. Some of these presented evidence which was open to doubt for a variety of reasons; what is more relevant is that it involved, throughout the 25 years, less than 50 pigs. On the other hand, in a recently published report on their extensive world survey, UKASTA produced an estimate indicating that the use of copper as an additive to diets of growing pigs resulted, in the UK, in a cost-benefit of £2.75/pig compared to when no copper was added to the diet. On this basis one can estimate that pig producers in the UK have benefited to date to the extent of about £600 million, and producers in the world by several billion pounds sterling. I am quoting these astronomical figures with one purpose in mind, namely to suggest that it is high time that the controversy on copper as a feed additive for pigs should be put in correct perspective; too many people have been approaching this subject rather superficially, with too little knowledge and understanding of all the factors involved, and often reacting to results from small, inadequate experiments, or just following hunches and speculations to arrive at unwarranted conclusions which do nobody any good, but could destroy the 'golden egg' already laid. It is important to insist on scientifically sound evidence and avoid emotional, often biased, intrepretation of incomplete evidence.

I would like to quote a sentence from my 1975 review: "Unfortunately, when one deals with important subjects affecting a large industry such as animal production, and particularly animal feeding, so many vested interests are involved, that often the technical scientific evidence is obscured, or sometimes intentionally or unintentionally veiled, misquoted or misinterpreted."

Here, if I may, I would like to strike a personal note: the whole 'copper story' followed an observation which I made about 35 years ago which indicated that pigs have a craving for copper. Eventually, this led to the findings that relatively high levels of supplementation (250 mg Cu/kg diet) exert a beneficial effect on performance. Ever since, I have endeavoured to assemble and study all the relevant scientific and technical facts which could guide us on the subject. I never had any vested interests in the 'copper story' and regrettably I did not succeed in unravelling the secrets of how copper exerts its performance promoting effect on pigs.

-4-

Copper is an essential element for life, but our under-standing of its functions, activities and reactions, for good or evil, is still inadequate. One can reiterate the fact that copper can be a toxic element, but at the same time one should not forget that it is well established in biology that a substance may be toxic at one level of administration and life saving or beneficial at another (e.g. arsenic, strychnine, aspirin, etc.). I can recommend very strongly a recently published, very comprehensive, review by the late Dr. K.E. Mason (1979) on 'A conspectus of research on copper metabolism and requirements of man'. (Incidentally, it has 879 references with not even one mentioning copper as a performance promoting feed additive). This review makes it abundantly clear how complicated and often very awkward a substance copper is, and perhaps it may warn some people from rushing to hasty conclusions and generalisations on meagre or circumstantial evidence. One should also keep in mind that recent evidence points in the direction that application of copper and its alloys may have a great potential for large scale marine acquaculture, and as a supplement to fertilisers used in cereal production.

In this introduction, I wish to quote a paragraph from my paper on 'Growth-promoting substances' presented at a Symposium of the Nutrition Society (Braude, 1976):

"As with many biological variables, response to a growth-promoting additive in a population usually follows a normal distribution. Once this distribution has been established, a further single record, however well established, can do no more than provide one point on the distribution curve. This should be only too obvious, but frequently claims are based on a single experiment, even if it contradicts a well established response. Often the culpable author shelters behind the statement that the results of his experiment were 'statistically significant'. With growth-promoting substances, the response to which is often affected by

many interacting and sometimes antagonistic factors,
one should not be allowed to challenge the established
distribution without a very substantial replication.
I will illustrate this point with evidence concerning
copper sulphate as a growth-promoting additive for
diets of growing pigs. Recently, I have reviewed this
subject (Braude, 1975). The results for improvement of
growth by addition of 250 mg Cu/kg diet when compared
with performance of the control animals is shown
diagrammatically in Figure 1. The mean response of
9.1% was statistically highly significant and the two
extreme values are also given. One can clearly see
that an additional single point, or even several points
added to the diagram, would not substantially alter the
conclusion that addition of 250 mg Cu/kg diet has a
beneficial effect on growth".

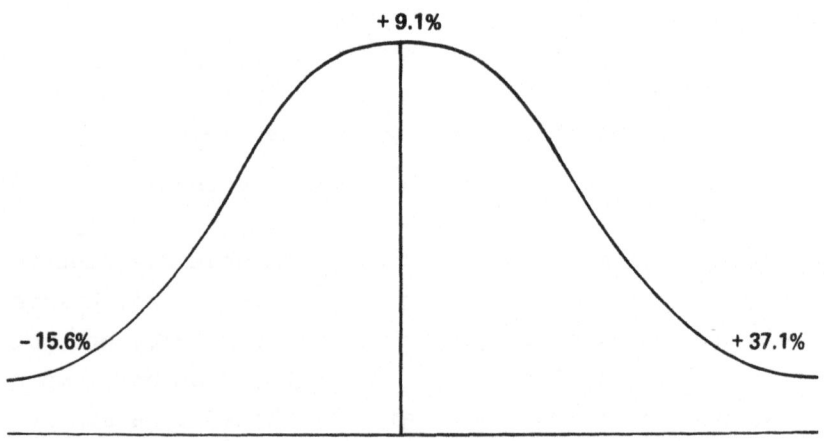

Fig. 1. The improvement of growth in pigs by addition of 250 mg copper/kg
diet.

The diagrammatically presented data were taken from 119 experiments carried out during the ten year period 1965 - 1975. During the earlier decade, 1955 - 1965, 83 experiments gave a very similar average value for improvement in daily live weight (DLG) gain - 8.1 v 9.1%. Since 1975, data from a further 43 experiments have been published involving supplementation with 175 - 250 mg Cu/kg diet. It is rather gratifying to find that all the additional evidence published since my two reviews (Braude, 1967 and 1975) augments the earlier evidence and fully supports the conclusions drawn that copper supplementation improves the performance of growing pigs. Additional evidence on issues that caused some anxiety has helped to clear some of the misgivings and the overall picture is reassuring.

In the next part of this paper I intend to quote briefly results of five experiments completed within the last two years by my colleagues and myself (Braude and Hoskins, 1980; Barber, Braude and Mitchell, 1980 - submitted for publication). The object of this presentation is to illustrate, once again, that with a difficult subject like copper as a feed additive, in which so many interactions are involved, it may not be possible to resolve outstanding problems by carrying out large scale, well planned and executed experiments. Only by critical assessment of all the accumulated evidence can one be reasonably guided, but as it is far from complete, one should neither request or require guarantees of 100% safety, which cannot be given.

In Tables 1 - 4 some results are summarised of three experiments completed within the last year. The experiments were designed to supply additional information to guide discussions and negotiations with the EEC, and particularly to determine whether any useful purpose could be served by reducing dietary copper supplement in the latter stages of the growing period. Suggestions, and indeed claims, were made (without adequate evidence) that by reducing dietary copper to 125 ppm at about 35 kg liveweight the growth promoting benefits could be maintained while liver storage of copper and also the

total copper excretions could be considerably reduced. The
latter point is being pressed by people concerned with
environmental pollution.

Experiment 1, Table 1, was carried out at NIRD and
produced evidence that addition of 200 ppm Cu _throughout_ the
growing period benefited performance most. Reducing the
dietary Cu to 125 ppm at 54 kg liveweight virtually annulled
the benefits which accrued in the earlier period. There was
no advantage from increasing the Cu level to 250 ppm during
the first period.

TABLE 1

RESPONSE TO 200 AND 250 mg Cu/kg DIET AND TO REDUCTION TO 125 mg DURING
THE LATTER PART OF THE GROWING PERIOD (EXPERIMENT AT NIRD)

24 pigs per treatment

Supplementary copper in diet (mg/kg)						
Treatment number	1	2	3	4		
Start to 54 kg lw	0	200	200	250	SEM	
54 kg to slaughter	0	200	125	125	(67 df)	
						Range test*
Daily gain, g						
Start to 54 kg	607	641	618	634	12.33	1 3 4 2
54 kg to 90 kg	796	813	789	795	12.07	3 4 1 2
Start to slaughter	674	701	679	691	10.17	1 3 4 2
Feed:gain						
Start to 54 kg	2.81	2.66	2.70	2.68	0.03	2 4 3 1
54 kg to 90 kg	3.52	3.51	3.64	3.54	0.055	2 1 4 3
Start to slaughter	3.11	3.01	3.11	3.05	0.031	2 4 3 1

*   Treatments underlined by same line not significantly different at
    P = 0.05

Experiment 2, a large scale co-ordinated trial with 25 centres participating designated as Expt. ARC 19, is reported in Table 2. Five treatments were involved. Due to an error in preparation of the experimental meal mixtures, the diets contained higher levels of Cu than intended. The first part of the test was completed (up to 54 kg liveweight), and the second part was abandoned.

The striking result of this test was that the highest level of Cu (about 275 ppm) produced the best performance.

TABLE 2

RESPONSE TO 200 and 250 mg Cu/kg DIET

Treatment number	1	2	3	4	5	SE (56 df)
Supplementary Cu, mg/kg proposed*						
up to 54 kg liveweight	0	125	200	200	250	
from 54 kg liveweight	0	125	200	125	125	
Actual Cu, up to 54 kg liveweight	0	160	230	230	275	
Daily gain, g	537	571	581	568	603	7.3
Feed:gain	2.86	2.69	2.66	2.72	2.55	0.032

*See text

An interesting and educational point arose from the error involved in the preparation of the experimental diets. It concerns the checking of levels of copper or indeed other feed additives in compound mixtures. Thirty samples were taken from a 30 tonnes consignment of a diet which was supposed to contain 250 mg of supplementary Cu/kg. The samples were taken during mechanical, automated bagging after delivery of every tonne. Each sample was analysed in triplicate and the results are presented in Table 3. It is quite obvious that one has to be very careful not to attach undue importance to analysis of single samples of multi-ingredient feed mixtures.

TABLE 3

SAMPLING OF 30 TONNES OF DIET FOR COPPER CONTENT 30 SAMPLES TAKEN AT 1 TONNE RUN-OFF mg Cu/kg DIET

Intended: 250 + 10 = 260

Mean for 30   252 within 3% of intended

		No.
(247-273 within	5%	7)
(234-288 within	10%	15)
208-312 within	20%	24
202-3		3
148		1
324-7		2

Experiment 3, the main co-ordinated trial, ARC 19A, is reported in Table 4. The results indicated a beneficial effect of Cu at 125 ppm level of supplementation being as good as that recorded at 200 or 250 ppm Cu, and that no adverse effects could be attributed to lowering the Cu level to 125 ppm after the pigs reached 54 kg liveweight.

There were no proven differences between treatment means for any of the carcass measurements except that the mean thickness of fat over the eye muscle for treatment 3 was greater than for treatments 2 and 5 ($P < 0.05$).

In Table 4 the concentrations of copper in the liver are reported. These were determined for pigs from only three centres. The mean overall growth rates for these pigs were 587, 621, 598, 609 and 615 g/day on treatments 1 - 5 respectively, and any differences between them were not proven ($P < 0.05$). The mean concentration of copper in the liver was significantly higher for treatment 3 than for any of the other treatments, and higher for treatments 5 and 4 than for treatment 1, whilst the mean concentration for treatment 2 was not proven different from either treatments 1 or 4.

The results of the three experiments have, unfortunately, not resolved the issues they were designed to clarify and only focussed attention on the need for great care when interpreting results of single experiments.

TABLE 4

RESPONSE TO 200 AND 250 mg Cu/kg DIET AND TO REDUCTION TO 125 mg DURING THE LATTER PART OF THE GROWING PERIOD

ARC Co-ordinated Test No. 19A carried out in 23 centres

	Supplementary copper in diet (mg/kg)					
Start to 54 kg	0	125	200	200	250	SE
54 kg to 90 kg	0	125	200	125	125	(65 df)
Treatment	1	2	3	4	5	
Daily gain						
Start to 54 kg	559	559	575	566	569	6.2
54 kg to slaughter	683	753	711	733	751	10.7
Start to slaughter	612	639	631	632	642	5.4
Feed:gain						
Start to 54 kg	2.80	2.79	2.75	2.81	2.78	0.023
54 kg to slaughter	3.84	3.52	3.72	3.61	3.55	0.051
Start to slaughter	3.30	3.15	3.21	3.19	3.15	0.026
Copper in liver, mg/kg DM	10.9	15.2	40.3	21.0	25.4	0.048

In Tables 5 and 6, some results are presented of two other experiments in which response to copper supplementation was compared on cereal diets containing, as a protein supplement, either fishmeal or soya bean meal. Earlier work, at the same centre (Barber et al., 1962), indicated that the addition of 250 mg Cu/kg diet produced a significantly better performance when fishmeal was used. On this occasion, no difference of any significance was found between the two diets with different sources of protein, but there was hardly any response to copper supplementation.

TABLE 5

FISH MEAL v SOYA BEAN MEAL AS PROTEIN SUPPLEMENT WITH AND WITHOUT 250 mg Cu/kg DIET

12 pigs/treatment - individual feeding

20 - 90 kg liveweight

						Mean effects	
					SE of means	Protein	Copper
Treatment number	1	2	3	4	(32 df)	supplement	sulphate
Protein supplement	Fish meal		Soya bean meal				
$CuSO_4 5H_2O$, 1000 mg/kg	–	+	–	+			
Daily gain, g	624	640	619	654	12.2	–4	26*
Feed:gain	3.51	3.44	3.56	3.58	0.053	0.00	–0.12*
Cu in liver, mg/kg DM	48	706	42	388		0.16**	1.07***

* 0.05 > P > 0.01

** 0.01 > P > 0.01

*** P < 0.001

† Analysis carried out on logarithmic values

TABLE 6

FISH MEAL v SOYA BEAN MEAL WITH 250 mg Cu/kg DIET

10 pens of 6-7 pigs/treatment - group feeding

20-90 kg liveweight

	Treatment number			SE	Multiple† range test
	1	2	3		
	Fish meal	Soya bean meal			
$CuSO_4 .5H_2O$, 1000 mg/kg	+				
Daily gain, g	628	602	618	4.074	2 <u>3 1</u>
Feed:gain	3.14	3.27	3.19	0.016	2 3 1
Cu content in liver, mg/kg DM	308	53	185		2 3 1

Differences between treatments underlined with the same line are not significant, P > 0.05

By studying carefully these six tables and comparing results between and within the experiments, one becomes aware of the great variation in responses to copper supplementation. Having often been confronted personally by results of this kind I cannot recommend too strongly that in matters concerning the use of copper as performance promoter for growing pigs, one should exercise the greatest possible constraint and certainly avoid generalisation based on inadequate evidence. One should not overlook the fact that about 15% (1 in 6) of pigs do not respond to copper supplementation of diets. We do not know why, but after all, we do not know why 85% do.

As this meeting concerns itself with the effect on the environment of copper as a feed additive to the diets of pigs, and as experts in this field are around the table, I have selected a few relevant comments which, as an interested layman, I hope you will find helpful.

I wish to draw attention to the views of Professor Todd from Northern Ireland who expressed concern about the occurrence of copper deficiency in ruminants in N. Ireland which he attributes to changes in the usage of copper compounds in agriculture during this century. He pointed out that reduction in potato growing and the replacement of copper based fungicides by non-copper preparations have resulted in a major reduction in copper input to agriculture in N. Ireland. This was further accentuated by replacement of copper sulphate by other anthelmintics. Professor Todd estimated that current use of copper sulphate in pig feeding represented only one-tenth of that formerly used within agriculture.

Our knowledge of the fate of the copper in pig excreta is still very limited. Not only are we uncertain in what form copper is found in faeces of pigs, what complexes it forms and how dynamic these processes are, but we are often puzzled and sometimes intrigued by the numerous reports on the effect of copper on different soils and plants. I wish to draw attention to three recent reports published under the auspices

of the International Copper Research Association: No. 269
(1979) by McLarren, Williams and Swift of the Edinburgh School
of Agriculture, UK on 'Studies on factors affecting the
availability of soil copper to plants', and Nos. 292 (1979)
and 292(A) (1980) by Martens and his colleagues of the Virginia
Polytechnic, Blacksburg, USA on 'Field experiments to evaluate
the plant availability of copper in pig manure'. The emerging
picture is by no means clear cut, and I will be looking forward
to the evidence which will be presented at this meeting, to
help to elucidate these problems.

However, as this meeting is convened by the EEC, I
cannot refrain from mentioning an amusing but nevertheless
disturbing development concerning copper, which points to a
rather bureaucratic approach to the copper problems and
certainly makes the scientist's mind boggle! Amid the
controversy whether the entry of 200 - 250 mg Cu/kg diet should
be transferred from Annex II (which allows 200 mg Cu/kg diet)
to Annex I (which allows 125 mg Cu/kg) an EEC directive was
released by a different department enforcing the use of copper
sulphate as a 'marking' ingredient for skim milk before it can
be used at a subsidised price for incorporation into animal
feed. At the level recommended (200 g/1000 l) the pig would,
in many circumstances receive more copper than it would receive
from supplementation of its meal mixture with 250 mg Cu/kg diet.

To end, I will summarise my views as follows: At this
moment in time, and knowing most of the existing evidence and
all the arguments for and against the use of 200 - 250 mg Cu/kg
diet, I see no reason why such a practice should not be allowed
to continue. Experience of the last 25 years would certainly
support this conclusion. Those who still have some misgivings
should accept the use of copper as an additive to diets for
growing pigs on the understanding that it could be stopped
overnight by administrative action, should vigilant and
continuous monitoring indicate that any undue hazard is
involved.

REFERENCES

Barber, R.S., Braude, R. and Mitchell, K.G., 1962. Copper sulphate and
  molasses distillers dried solubles as dietary supplements for
  growing pigs. Anim. Prod., 4, 233-8.

Barber, R.S., Braude, R., Mitchell, K.G. and Cassidy, J., 1955. High
  copper mineral mixture for fattening pigs. Chem. & Ind., 601.

Braude, R., 1967. Copper as a growth stimulant in pigs (*cuprum pro
  pecunia*). Trans. Symp. '*Cuprum pro vita*' Vienna p.55 and Wld. Rev.
  Anim. Prod., 1967, 3(2), 69-83.

Braude, R., 1975. Copper as a performance promoter in pigs. Trans. Symp.
  Copper in Farming, London, 79-94.

Braude, R., 1976. Growth promoting substances. Proc. Nutr. Soc., 35,
  377-382.

Mason, K.L., 1979. A conspectus of research on copper metabolism and
  requirements of man. J. Nutrit., 109, 1979-2066.

Meyer, H. and Kröger, H., 1973. Copper in pig diets. Ubers. Tierernähr.
  1, 9-44.

Omole, T.A., 1980. Copper in the nutrition of pigs and rabbits: a review.
  Livestock Prod. Sci., 7, 253-268.

UKASTA, 1978. Survey relating to copper in pig feeds. United Kingdom
  Agricultural Supply Trade Association, 1-93.

Wilson, P.N., Brigstocke, T.D.A. and Cooke, B.C., 1979. Copper as an
  inexpensive growth promoter for the pig. Process Biochemistry
  published by Wheatland Journals Ltd, Watford, Herts, 1-4.

PAST AND PRESENT SITUATION IN RELATION TO THE USE OF FEED ADDITIVES IN DIETS FOR PIGLETS; CONSEQUENCES OF UTILISATION OF COPPER

A. Aumaitre*

Pig Husbandry Department, Centre de Rennes-St-Gilles,
35590 L'Hermitage, France.

ABSTRACT

Growth promoters or antibacterial agents are particularly efficient when they are used as feed additives for rearing early weaned piglets. Improvement in performance includes reduction of mortality rate, improvement in daily growth rate and decrease in food conversion ratio. Numerous products might be used and the maximum authorised levels of antibiotics, organic acids or salts, as well as organic compounds are listed as approved by the specialist committee of the EEC.

Experience has shown that high levels of copper (250 ppm) are favourable for growing finishing pigs but they appear to be inconsistently efficient for piglets up to 20 kg. Restriction to less than 100 ppm would appear reasonable for starter diets. Consequences of utilisation of large amounts of copper in pig diets are considered by measuring the residual amount contained in pig slurry generally spread as fertiliser.

Advice and recommendations on the choice and use of starters supplemented with additives are given, based on consideration of their respective efficiency in controlled experimental trials.

*with the cooperation of J.P. Raynaud, B.P. 42, 37400 Amboise, France.

INTRODUCTION

In the 1950s, considerable attention was paid to the possibility of using antibiotics in the diets of pigs and particularly in the diets of suckled and weaned piglets. Numerous experiments throughout the world were designed and carried out, and the results were published in the literature (Hays, 1969).

Nevertheless, when the results were considered in more detail, considerable variations in experimental design or procedure were observed:
- the age of the animals involved in the experiment
- the duration of the treatment (2 weeks - 6 months)
- the kinds of feed additives, their biological properties and the levels used
- the number and quality (health) of the animals at the beginning of the experiment.

In France, experiments were conducted to measure the efficiency of antibiotics on the one hand and of feed additives, including antibacterial agents and copper on the other. These experiments were measuring the effects on:
- the animals' growth, food conversion ratio, scouring conditions and mortality rate (Février, 1958; Salmon-Legagneur, 1961; Aumaitre, 1968)
- the metabolism of the animals (Francois, 1962; Ferrando, 1975)

The aim of this report is to summarise some of the most important results and their consequences in the practical feeding of piglets.

Feed manufacturers and pig breeders are very sensitive to the efficiency of new feed additives. Everybody trusts to the utilisation of feed additives for the solution of scouring mortality and morbidity problems rather than to the improvement of general management conditions or hygiene. For this reason, the utilisation of additives in piglet diets is quite common.

Despite strict EEC regulations concerning the use of feed additives, it appears that two main problems have not been definitively solved:
- are the additives used at the optimum level of efficiency for animals of different ages?
- does chronic use, even at recommended levels, of additives have a deleterious effect on the environment (residues in meat and wastes) ?

These points will be discussed in the present paper by considering the benefits or otherwise of growth promoters to piglets.

1. DIFFERENT FEED ADDITIVES AUTHORISED FOR PIGLET DIETS

Numerous growth promoters, generally added to diets at a relatively low level, have been discovered in the past 30 years (Hays, 1969). Their efficiency has been attributed to their antibacterial properties as well as to their effect on protein and energy metabolism (Francois, 1962). Because of the possible use in human or animal therapeutics, continuous supplementation of piglet feeds is subject to strict controls (Ferrando, 1975, Table 1).

Furthermore, the efficiency of feed additives can be classified according to their particular biological properties (Table 2). At the approved level of supplementation, the product can be a growth promotant and/or a product for prevention of digestive disturbances.

TABLE 1

FEED ADDITIVES AUTHORISED FOR CONTINUOUS SUPPLEMENTATION OF PIG DIETS
(FRANCE - EEC, 1980)

		Pig Diet		Maximum age
		ppm mini	ppm maxi	(or withdrawal)
	Polypeptides			
	Bacitracine			
	Annex 1	5	20	6 months
	Annex 2	20	50	10 weeks
	Flavomycine			
	Annex 1	1	20	6 months
	Avoparcine			
	Annex 1	5	20	6 months
	Annex 2	10	40	10 weeks
	Macrolides			
Antibiotics —	Oleandomycine			
	Annex 1	2	10	6 months
	Annex 2	10	25	10 weeks
?	Spiramycine			
	Annex 1	5	20	6 months
	Annex 2	20	50	
	Virginiamycine			
	Annex 1	5	20	6 months
	Annex 2	20	50	10 weeks
—	Erythromycine			
	Annex 2	5	20	
?	Tylosine			
	Annex 1	5	20	2 - 6 months
	Annex 2	10	40	2 months
	Furannes			
—	Nitro-Imidazole			
	Dimetridazole			
Medicated	Annex 2	100	200	30 days before slaughter
Substances				
—	Ronidazole			
	Annex 2	30	60	30 days before slaughter
	Furannes			
	Nitrovine			
	Annex 1	5	15	6 months
	Annex 2	10	25	10 weeks
Growth	Quinoxaline			
Promoters				
	Di.N Oxides			
	Carbadox			
	Annex 2	10	50	4 weeks before slaughter
	Olaquindox		100	

- medicated feed only ? withdrawal end of 1980   Aumaitre and Raynaud, 1978.

TABLE 2

EFFICACY OF FEED ADDITIVES AT THE MAXIMUM AUTHORISED LEVEL, OR AT HIGH LEVELS

	Dose ppm	Maximum approved dose		Dose level which could be preventive
		Is the product a growth promotant?	Is the product preventive at this dose?	
Bacitracin	50	Yes	No	?
Flavomycin	20	Yes	No	?
Avoparcin	40	Yes	No	?
− Oleandomycin	25	Yes	No	200 ppm
? Spiramycin	50	Yes	No	200 ppm
Virginiamycin	50	Yes	Swine dysentery	same or >
− Erythromycin	20	Yes	No	200 ppm
? Tylosin	40	Yes	No	200 ppm
Furazolidone	100	Yes ±	No E. coli Salmonella	same or >
− Dimetridazole	200	Yes ±	S.D.+Enteritis	same or >
− Ronidazole	60	Yes ±	S.D.	same or >
Nitrovin	25	Yes	No	50 - 100 ppm
Carbadox	50	Yes	S.D. E.coli Salmonella	−
Olaquindox	100	Yes	S.D. E. coli	same or >

S.D.: Swine dysentery

"Yes" for growth promotant means products known as growth promotants *stricto sensu*

± means products that have more antibacterial properties

(Aumaitre and Raynaud, 1978)

2. THE ROLE OF FEED ADDITIVES IN REARING OF PIGLETS

2.1. Effect on growth, feed efficiency and mortality

The use of antibiotic substances was very common in the 1950s. Very low levels of substances were recognised as being efficient in preventing mortality of runt piglets generally weighing less than 10 kg at 8 weeks (Table 3). At the same time the average daily gain of piglets was increased and the food conversion ratio was decreased for animals up to 30 kg.

TABLE 3

EFFICIENCY OF ANTIBIOTIC SUPPLEMENTS ON THE SURVIVAL RATE OF RUNT PIGLETS AND ON THEIR PERFORMANCE

		Treatment (1)	
	Control	Penicillin 20 ppm	Chlortetra-cycline 10 ppm
Average daily gain g	266		
% Improvement		14 (2)	23 (2)
Food conversion ratio	3.9		
% Improvement		9	15
Mortality rate %	27	22	8

(1) 8 - 30 kg
(2) calculated for surviving animals

The average improvement in performance is generally observed with a supplement of antibiotics or antibacterial products added and mixed to the feed.

Numerous observations made from experiments published in world literature provided the following evidence:

1. The importance (or the percentage) of the growth improvement is inversely proportional to the performance of the control animals (Figure 1)

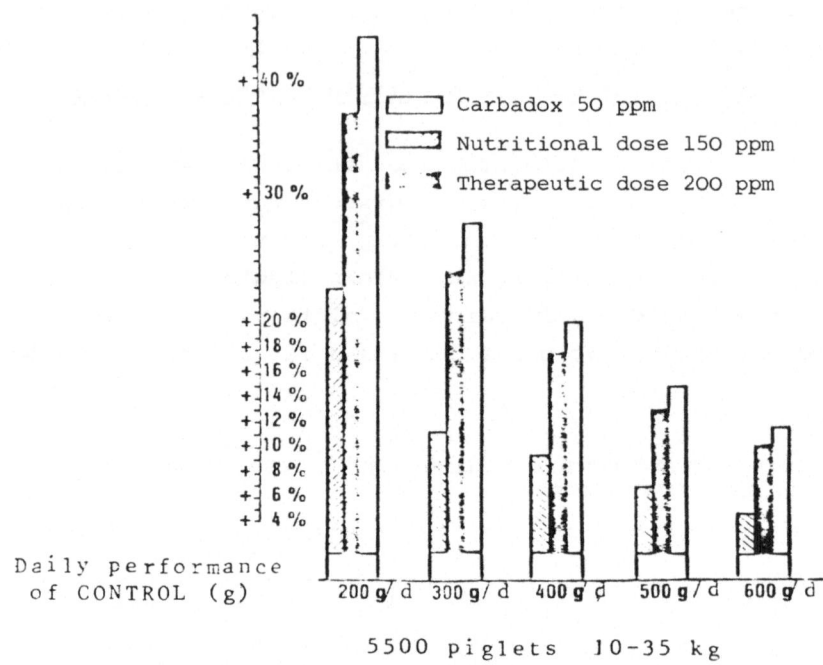

Fig.1. Improvement in average daily gain of piglets due to feed additive
utilisation in relation to the performance of control animals
(Raynaud, 1974)

2.   For accurate levels of supplementation, the growth response
      of the animals is improved linearly by increasing the
      dose of the active substance added to the feed (Figure 2).
      Nevertheless, the maximum improvement of 30 - 40% in the
      average daily gain was observed for runt animals affected
      by severe diarrhoea or raised in very poor conditions of
      hygiene and environment.

      One of the most important effects of feed additives in
the diets of piglets consisted of a significant improvement of
the average food intake; this was generally observed in all
experiments described in the literature (Table 4).

TABLE 4

IMPROVEMENT IN PERFORMANCE OF PIGLETS DUE TO USE OF FEED ADDITIVES (EXPRESSED AS A PERCENTAGE OF IMPROVEMENT IN COMPARISON WITH CONTROL ANIMALS BETWEEN 7 AND 25 kg)

Additive	Oxytetra-cycline	Oleandomycine	Oxytetracycline + oleandomycine	Tylosine
Level ppm	100	20	40 + 10	60
Average daily gain %	11	29	35	27 (+9 +50)
Food intake %	7	17	45	16 (+8 +48)
Food conversion %	-3	-10	+7	7 (-15 +4)
Authors		Aumaitre 1968		Vanschoubroek and de Wilde 1965

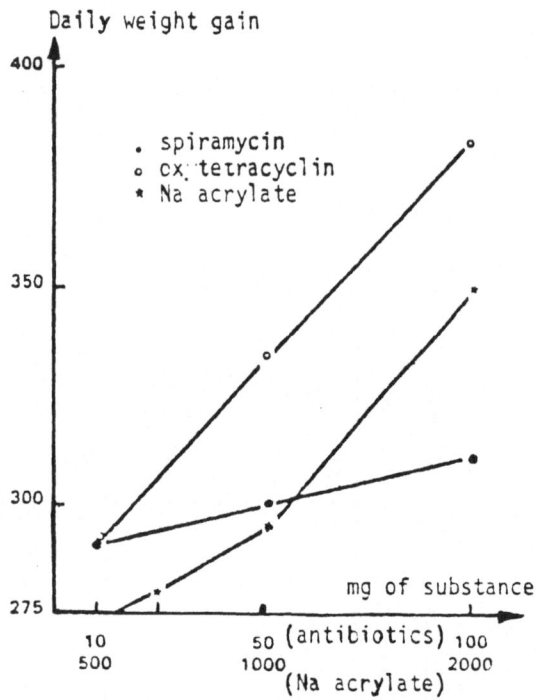

Fig. 2. Growth response to various doses of each substance given to
piglets after weaning (5 weeks old). (Jouandet et al., 1964)

The maximum improvement in food consumption, as
well as in daily gain of the animals, would be associated, in
borderline cases, with an increase in the food conversion ratio.
However, in general, an improvement in the health conditions
of the piglets by decreasing scouring frequency is related to
an increase in the food intake.

Further experimental evidence of the additional beneficial metabolic effect of feed additive supplementation has
been demonstrated in piglets (Table 5).

In an experimental pair-feeding of young animals a
significant improvement in growth rate, and consequently in
food conversion ratio, was found in our experiments as well as
in that of Brown, Becker et al. (1952)

TABLE 5

FEEDING RESPONSE OF PIGLETS TO DIET SUPPLEMENTATION IN A PAIR-FEEDING
EXPERIMENT

Authors	Brown et al., 1952		Aumaitre, 1963	
Animals	16 - 45 kg		7 - 20 kg	
Supplement ppm	0	25 CTC (1)	0	40 OTC (2) +TO oleando
Food intake g/day	1253	1253	423	424
Improvement in average daily gain	-	+3%	-	+5%

(1) CTC : aureomycine   OTC : terramycine
(2) TO : terramycine + oleandomycine

Inconsistent results were observed with the use of weak
organic acids, or their salts, as feed additives (Table 6) for
piglets as well as for growing finishing animals. Furthermore,
products such as salicylic acid acetate, already used as
chemotherapeutic drugs for humans or animals, would never be
allowed in animal feeds by the medical authorities.

2.2. Effect of environment on the efficiency of feed additives

In the past 20 years there has been a considerable increase in the size of pig herds in Europe. An increase in the
number of animals in the same building is associated with a
reduction of the floor space allowed to each piglet. As a
consequence of an increase in labour costs, the use of concrete
floors has become general practice for growing finishing pigs
and, unfortunately, for sows and young piglets as well.

Feed additives appeared to be more efficient for piglets
reared in bad housing conditions, particularly for animals
raised in humid conditions on poorly insulated concrete floors
in comparison with animals raised on a 'flat deck' (Table 7).

TABLE 6

EFFICACY OF WEAK ACIDS GIVEN AS FEED ADDITIVES ON THE PERFORMANCE OF YOUNG OR GROWING SWINE
(AUMAITRE AND RAYNAUD, 1974)

Substance	Swine live weight (kg)	Dose	Improvement of performances in %		Reference
			weight gain	feed/gain	
	5 - 19	0.75 %	+ 11	+ 2	van Kempen et al. 1969
		0.75 %	0	- 3.4	van Kempen et al. 1970
		0.75 %+ 10 ppm antibiotics	0	0	
Citric acid	5 - 20	0.5 %	+ 1	0	Kirchgessner and Roth-Maier, 1975
		1.5 %	- 2	+ 2	
		4.5 %	+ 8	0	
	20 - 100	0.7 % +10 ppm antibiotics	+ 4.5	+ 1.5	Devuyst et al., 1973
		0.7 %	0	0	Devuyst et al., 1974
Fumaric acid	7 - 23	0.5 %	- 8	+ 2	Kirchgessner and Roth, 1975
		1 %	0	+ 4	
		2 %	+ 12	+ 7	
		4 %	+ 4	+ 5	

(Contd.)

TABLE 6 (contd.)

Substance	Swine live weight (kg)	Dose	Improvement of performances in % weight gain	feed/ gain	Reference
Sodium salicylate	8 - 20	500 ppm	+ 9	+ 7.5	Aumaitre and Rerat. 1971
	22 - 90	200 ppm	+ 0.6	0	1971
		700 ppm	+ 4	+ 6	Barber and Braude, 1974
Salicylic acid acetate	8 - 20	500 ppm	+ 37	+ 12	Aumaitre and Rerat, 1971
Sodium acrylate	8 - 20	1000 ppm	+ 5	+ 5	Jouandet et al., 1964
	8 - 20	2000 ppm	+ 17 to 24	+ 12 to + 10	Michel et al., 1964

TABLE 7

EFFECT OF VARIOUS HOUSING CONDITIONS ON THE EFFICIENCY OF CARBADOX OR OLAQUINDOX FOR IMPROVING THE PERFORMANCE OF PIGLETS

Pens	Concrete, no straw		Wire mesh (flat deck)	
Liveweight	10 to 20 kg	5 to 20 kg	10 to 20 kg	5 to 20 kg
(1) Schneider and Bronsch 1974				
D W G (g)				
Controls		403		436
% improvement Carbadox 50 ppm		+ 21%		+ 8%
Days diarrhoea/piglet				
Controls		1.7		1.4
Carbadox 50 ppm		0.4		0.8
(2) Bronsch et al., 1976				
D W G (g)				
Controls	469		482	
% improvement Olaquindox				
25 ppm	+ 14%		+ 11%	
50 ppm	+ 3%		+ 3%	
75 ppm	+ 14%		+ 7%	
100 ppm	+ 12%		+ 9%	

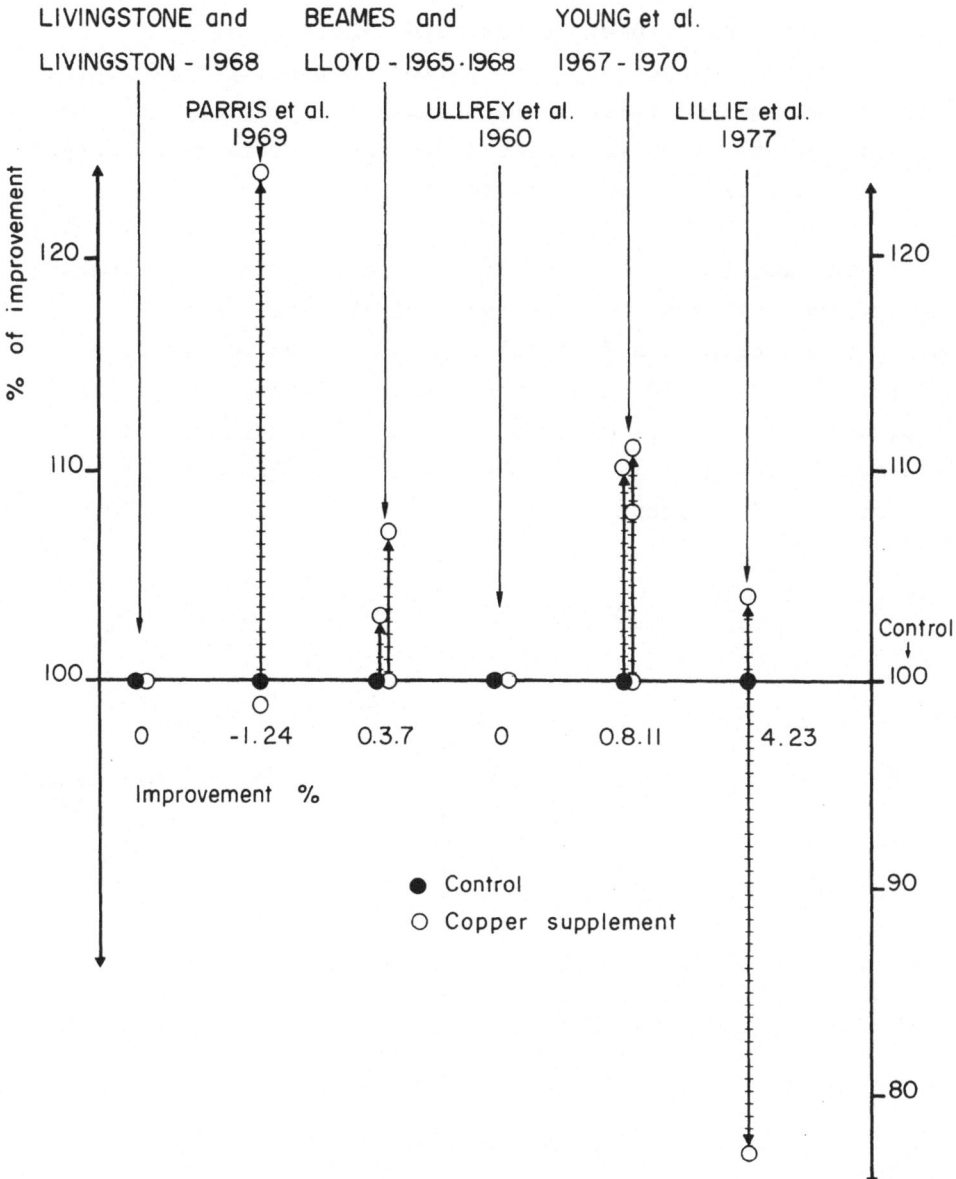

Fig.3. Different experimental responses of piglets to a copper supplement of 250 ppm in the diet. Improvement of growth rate.

2.3. Utilisation of copper in piglet diets

The efficiency of copper supplements for growing finishing pigs has long since been demonstrated (Braude, 1976). However, utilisation of copper supplementation in the diets of piglets was the subject of further investigation. Nevertheless, no definite conclusions were made about the efficiency of copper on piglets from 5 to 20 kg liveweight.

The addition of a high level of copper to a diet based on soyabean oil meal was apparently efficient; conversely, a similar addition to a fishmeal based diet was not efficient (Parris and McDonald, 1969; Young and Jamieson, 1970). In addition the presence of Cu^{++} activated the turnover rate of pig pepsin *in vitro* (Kirchgessner et al., 1976) but did not improve *in vivo* digestibility of protein (Beames and Lloyd, 1965).

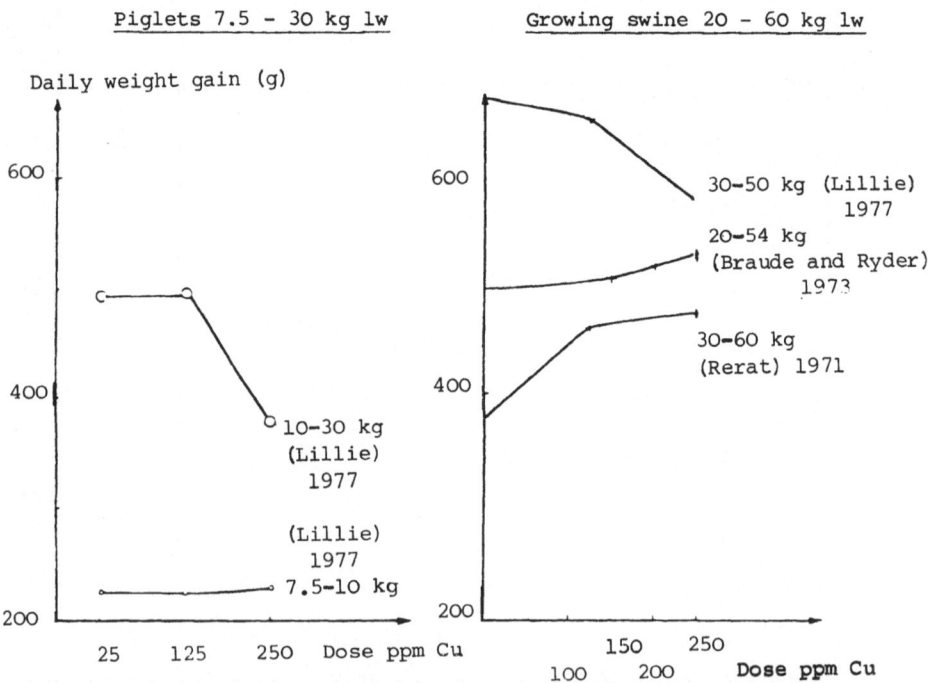

Fig. 4. Consequences on the growth of young or growing pigs of the introduction of high levels of copper sulphate into the diet.

TABLE 8

ANALYSIS OF TRACE ELEMENTS IN COMMERCIAL STARTERS FOR PIGLETS (AUMAITRE ET AL, 1979) ppm OR mg/kg ON THE AIR DRY BASIS

Trace element	Feed Age (1) Weight	ANALYSIS Mean	DATA Range	Requirement	Toxicity level
Magnesium	21 (5 - 10 kg) 35 (10 - 20 kg)	1 637 1 876	1 240 - 2 030 1 390 - 2 250	330 - 440 240 - 500	
Zinc	21 35	225 208	141 - 283 136 - 285	50 - 100	2 000
Manganese	21 35	91 95	40 - 134 30 - 144	12 - 40	4 000
Iron	21 35	543 420	243 - 954 275 - 588	80	5 000
Copper	21 35	118* 127**	64 - 160 40 - 203	6 (125 - 250?)	250 - 500

(1) Starter for weaning either at 21 or at 35 days
* ** 2 starters containing 10 ppm are not included in the mean

- 31 -

TABLE 9

ANALYSIS OF COPPER CONTENT IN THE SLURRY OF GROWING FINISHING PIGS REARED IN COMMERCIAL FARMS IN FRANCE (HEDUIT ET AL, 1977)

	THIS STUDY n = 26			KAHARI, 1974		FURRER, 1974
	Mean	V.C%	Range	Mean	Range	
MINERALS						
Ca	4.8	30	3.2 – 11	2.8	1.4 – 3.5	3.37
Mg % DM	1.5	20	0.8 – 2.16	0.7	0.5 – 1.3	0.891
Na	1.1	58	0.4 – 3	1.29	2.5 – 3.2	1.38
TRACE ELEMENTS						
Zn	1 120	19	720 – 1 750	345	96 – 864	1 597
Fe	2 620	23.4	1 520 – 4 243	999	450 – 1 290	1 686
Cu ppm in	838	29	348 – 1 365	418	96 – 846	249
Mn DM	576	34	280 – 1 200	248	123 – 392	415

V.C : variation coefficient between samples %

Furthermore, the depressive effect of a high amount of copper (250 ppm) was clearly demonstrated by Lillie et al. (1977) after a continuous supplementation provided at an early age. These indications should be taken into account in the definitive conclusions regarding recommendations for the utilisation of copper in the diets of piglets.

The level of supplementation in the diets of piglets was investigated in France by means of chemical analysis performed on commercial starters which were prepared by the feed industry. The average values were around 125 ppm, but large variations in the range were observed (Table 8). Higher levels were found when a unique premix of trace elements and vitamins was used for preparation of feed for pigs as well as for piglets.

The consequences of copper supplementation for growing finishing pigs were measured by analysis of pig slurry collected from 26 commercial farms (Table 9). Generally, large amounts of copper were found in the slurry dry matter as a consequence of the incorporation of a high level of copper into pig diets: up to 1 365 ppm were found in one sample. The total amount of copper spread in the fields, at the rate of 30 m^3/ha/year is around 4.1 kg Cu/year. This amount can be observed now, and is sometimes doubled in areas of western France which are specialised pig raising areas.

3. PRESENT SITUATION WITH REGARD TO THE UTILISATION OF FEED ADDITIVES

Three sorts of programmes could be used to improve performance and economic results in intensive animal production units (Figure 5).

3.1. Continuous low level supplementation

Until recently, this was the general method used with antibiotics and feed additives, but today the aim is for the level selected to be effective in preventing the occurrence of

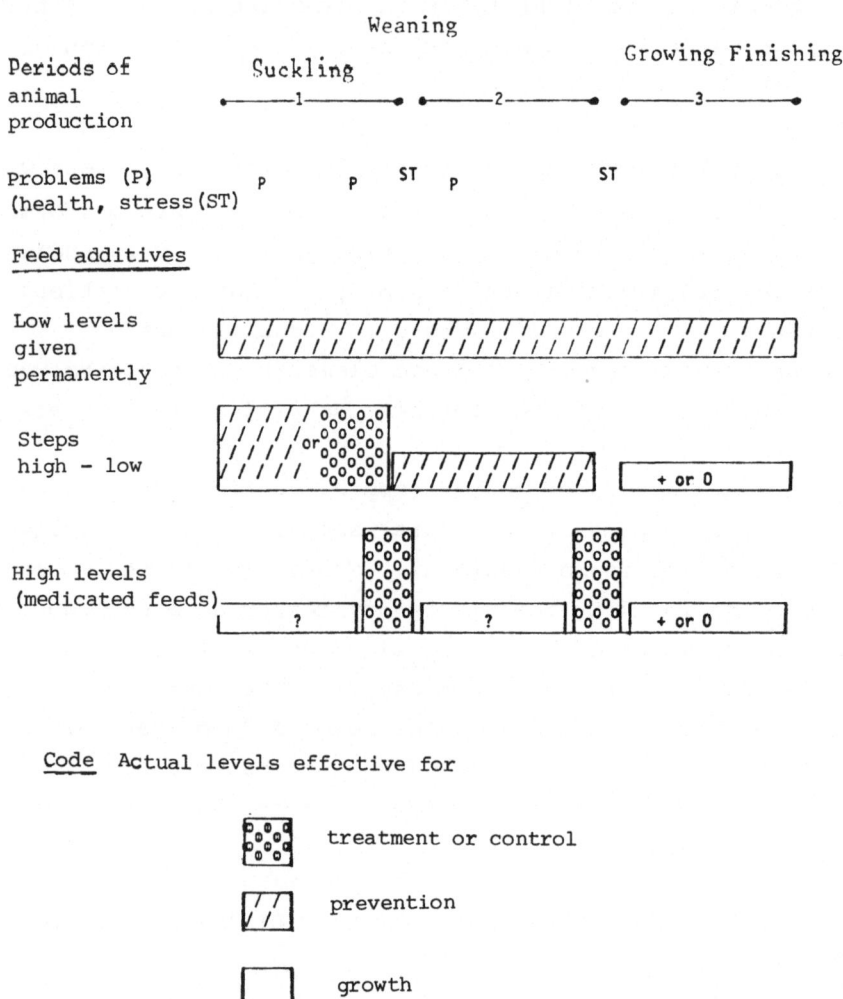

Fig. 5. Programmes for the use of feed additives or high levels of
medicated feed.

diseases and problems due to non-specific pathology.

This is the only way to optimise the performance and to avoid the need for curative treatments which are too late if they are performed once the problems have started.

3.2. Graduation

The level selected for the first period should be effective for the prevention or even the treatment of usual diseases which are economically important.

The second level should be effective only in the prevention of usual problems.

During the last period of growth, there was a choice as to whether

- to use a level which was only growth promoting
- or not to use any product, as is the case during the last (finishing) period.

3.3. High levels

Medicated feeds with high levels of antibiotics (generally in combination) are given under veterinary prescription at each critical period when stress is expected. In the intervening periods, a level of adjuvant could be used which has only growth promotant efficacy.

However, the first two programmes are better because they do not disturb the balance of micro-organisms in the intestinal flora. The use of high levels of antibiotics (third programme) is suspected of being responsible for disturbances of the intestinal flora and the possible transmission of multi-resistant micro-organisms to humans.

It is very important to know to which levels an additive

is constantly effective as a preventing agent, and against
which diseases it is effective. Table 1 indicates the efficacy,
at the highest approved level, as growth promotants *stricto sensu*
and as preventives, of products used as feed additives in
Common Market countries.

Ideally, a feed additive for piglets should fulfil the
following requirements:

1. In suckling piglets and pre- and post-weaning
    - suppress or considerably diminish mortality
    - prevent all diseases which could cause lesions and
      depreciate the carcasses at slaughter
    - prevent intestinal disturbances
    - suppress non-specific pathology, subnormal
      performance and the effects of stress
    - stimulate appetite

2. In young pigs up to 35 kg liveweight
    - prevent intestinal disturbances
    - suppress non-specific pathology, subnormal
      performance and the effects of stress
    - reduce heterogeneity of groups by helping 'poor
      performance' animals
    - stimulate appetite

3. In growing finishing pigs
    - stimulate growth and reduce the duration of the
      fattening period
    - improve the feed/gain ratio through a 'feed sparing
      effect
    - improve carcass quality

The present tendency to look for products which greatly stimulate growth and improve the feed/gain ratio at the start of production is to be encouraged. A low standard of hygiene or badly conducted weaning at the piglet stage can compromise the life and future of the animal, leading to the production of 'technological runts'. The use of additives should be particularly intended to do away with this type of situation which is caused by highly intensive production, mixing of animals or poor management.

REFERENCES

Aumaitre, A., 1968. Ann. Zootech. 17, 199-205.

Aumaitre, A. and Raynaud, J.P., 1978. Dossiers Elevage. 2, 73-83, 3, 47-77.

Aumaitre, A., Peiniau, J., Calmes, R. and Seguin, M., 1979. Jour. Rech. Porc. France, 11, 187-196.

Beames, R.M. and Lloyd, L.E., 1965. J. Anim. Sci. 24, 1020-1026.

Beames, R.M., 1969. J. Anim. Sci. 29, 573-580.

Braude, R., 1976. Proc. Nutr. Soc. 35, 377-382.

Bronsch, K., Schneider, D. and Rigal-Antonelli, F., 1976. Zeit. Tierphysiol. Tierernähr. Futtermittelkde. 36, 311-321.

Brown, Becker, D.E., Terrill, S.W. and Card, L.E., 1952. Arch. Biochem. Biophys. 41, 378-382.

Ferrando, R., 1975. World. Rev. Nutr. Diet. 22, 183-235.

Fevrier, R., 1958. Ann. Zootech. 6, 163-169.

Francois, A.C., 1962. World. Rev. Nutr. Diet. 3, 21-64.

Hays, V.W., 1969. The use of drugs in animal feeds. Pub. 1679. Nat. Acad. Sci. Washington. 11-30.

Heduit, M., Roustan, J.L., Aumaitre, A. and Seguin, M., 1977. Jour. Rech. Porc. 9, 305-310.

Jouandet, C., Aumaitre, A. and Salmon-Legagneur, E., 1964. Ann. Zootech. 13, H.S. 113-127.

Kirschgessner, M., 1976. Brit. J. Nutr. 36, 15-22.

Lillie, R.J., Frobisch, L.T., Steele, N.C. and Graber, G., 1977. J. Anim. Sci. 45, 100-107.

Livingstone, R.M. and Livingstone, D.M.S., 1968. J. Agri. Sci. Camb. 71, 419-424.

Parris, E.C.C. and McDonald, B.E., 1969. Can. J. Anim. Sci. <u>49</u>, 215-222.

Raynaud, J.P., 1974. Zeitsch. Tierphysiol. Tierernähr. Futtermittelkde. <u>19</u>, 309-320.

Schneider, D. and Bronsch, K., 1974. Züchtungskunde. <u>46</u>, 366-375.

Ullrey, D.E., Miller, E.R., Thompson, O.A., Zutaut, C.L., Schmidt, D.A., Ritchie, H.D., Hoefer, J.A. and Luecke, R.W., 1960. J. Anim. Sci. <u>19</u>, 1298.

Vanschouebroek, F.X. and De Wilde, R., 1965. Vlaams. Diergenees. Tigdsch. <u>38</u>, 213-240.

Young, L.G., 1967. J. Anim. Sci. <u>26</u>, 912 Abs.

Young, L.G. and Jamieson, J.D., 1970. Can. J. Anim. Sci. <u>50</u>, 727-733.

DISCUSSION

R. Braude (*UK*)

I would like to make a few comments by way of opening the
discussion. First of all, with regard to feeding copper to
piglets, we have never recommended feeding copper to piglets.
Our evidence shows that copper should be included from 20 kg
liveweight onwards. There is no firm evidence available on the
effect of feeding copper to pigs below this weight. As far as
the level for older pigs is concerned, in my view the evidence
is overwhelming that if you feed at 125 ppm you are wasting
your time because instead of one in six pigs not responding,
you will get two in six responding. The conditions in which
the pigs are kept is an important factor. I would not be
interested in copper as a feed additive if it was used to
replace good husbandry. I am only interested in the positive
effect of copper under the best conditions. Here you must be
very careful; if you are improving performance all the time and
getting nearer and nearer to the ceiling, then you cannot
expect to go on forever making improvements at the top. Under
some circumstances, a response of 2% would be very worthwhile.
I would like to quote some figures on copper in liver. These
are based on 48 reports covering different experimental
conditions and results, all summarised together. The mean for
control pigs was 36 mg Cu/kg DM. The range for the controls
was 9 - 170; this was for animals which had no copper added to
the diet. 250 mg copper added to the diet produced an average
of 708, ranging from 22 to 4 657. Except under the conditions
which produced 4 657 (which were very extreme) there were no
adverse effects observed in the pigs. What do these figures
mean? The high values in control animals and the low values in
the treated animals are puzzling.

K.L. Robinson (*UK*)

Dr. Braude mentioned the question of variability in
response to copper additions. He said that 15% of pigs do not
respond. Is he saying that that number of individual pigs in

any group do not respond, or is he saying that there is no
response in 15% of experiments? Secondly, I have seen figures
quoted which indicate that the higher the level of copper in
the diet, the less variability there is. This is given as an
argument in favour of 200 or 250 ppm copper in the diet in
contrast to 125 ppm. The variability is alleged to be less at
the higher level. Has Dr. Braude any comments on that point?

R. Braude

The figure of 15% is just an estimate from global assessment of the situation. Over the last 25 years we have done at
least 50 experiments with copper at our own Institute. Some
are included in the averages I mentioned; some have never been
published and therefore are not included. The same general
remarks I made on the global evidence apply to Shinfield
evidence. As far as variability on higher levels of copper is
concerned, that is true up to a point but one has to be careful in generalising on matters concerning the levels of copper
in the diet; it is often very difficult to establish what
levels were actually fed. Dr. Aumaitre referred to American
experiments, and there is also a big series of Canadian experiments where the researchers thought that they were feeding a
certain level of copper and, on scrutiny, found that quite
different levels of copper were actually fed.

D.B.R. Poole (*Ireland*)

In relation to the spread of values that Dr. Braude just
presented in terms of liver copper level, this does not
surprise me at all because what he is talking about is absolute
levels of copper regardless of other components in the feed.
It is quite clear from evidence on many species that the
availability of copper for absorption by the animal will depend
on the other components of the feed. In taking a wide spread
of experiments with different types of feed and different
components of zinc, protein and other factors, one would expect
a wide variation of absorption and therefore of liver copper.
I would like to ask Dr. Braude whether there is any relationship

between the effectiveness of copper as demonstrated by performance of pigs and the absorption of liver copper.

R. Braude

The answer is that there is no such relationship. I agree that the figures for total copper in liver which I gave can be meaningless unless you know what you are talking about but what is interesting is the question of how you can possibly get the figure of 170 in the controls and yet get the figure of 22 where copper was fed.

D.B.R. Poole

That is a positive statement. You did say earlier that there is no evidence for it.

R. Braude

The evidence says that there is no relationship. However, again you have to be very careful with these kinds of figures; comparing a post mortem result with what happened during lifetime could be misleading.

J.B. Ludvigsen (Denmark)

Dr. Braude, you did not give any copper concentration in the liver of the adult man but you said that in newborn babies it was ten times as high. Are these samples from the newborn babies taken when they are alive, or is it stillborn or dead babies? If the latter is the case, they could have died from copper poisoning?

R. Braude

That figure on babies comes from the recently published paper in the American Journal of Nutrition. I believe it related to live babies but I would have to check this.

DANISH EXPERIMENTS WITH COPPER FOR BACON PIGS

A. Madsen and V. Hansen

National Institute of Animal Science,
25 Rolighedsvej, DK-1958 Copenhagen V, Denmark.

ABSTRACT

In Danish experiments the maximum concentration of copper used as a growth promoting supplement to bacon pig diets has been 250 ppm Cu in the form of $CuSO_4 \cdot 5H_2O$. For the period 20 - 90 kg liveweight 125 ppm Cu improved daily gain and feed efficiency by 5%. No significant response was obtained in the period 50 - 90 kg. Accumulation of copper in the liver was low and may be diminished if the supplement is withdrawn after a liveweight of approximately 50 kg. This will also ensure a lower copper content in the pig manure. There was a tendency towards a slight increase in the iodine value of the backfat, when the diet was supplemented with copper, but otherwise no influence was observed on carcass quality.

INTRODUCTION

Copper has been used as a growth promoter in pig diets for many years in Europe, mainly in the UK, but is still not approved for use in the United States and Canada. The response of growing pigs to copper as a feed supplement has been extensively documented and a very comprehensive review was published by Braude (1975).

Since at this EEC workshop Dr. Braude will discuss results based on 25 years use of copper, this paper will only review Danish experiments with copper and discuss some practical aspects. Two series of feeding experiments with bacon pigs have been carried out in the years 1956 - 57 and 1973 - 74, respectively.

MATERIAL AND METHODS

The experiments comprised the period from approximately 20 to 90 kg liveweight. The pigs which were Danish Landrace, received either a control diet or the same diet supplemented with different levels of copper. Copper was added as copper sulphate ($CuSO_4.5H_2O$). Antibiotic was never added to the feed. The pigs were fed restrictedly according to a scale based on liveweight, and straw bedding was used.

Experiment 1956 - 57

This was conducted to compare supplements with different levels of copper ranging from 0 to 250 ppm Cu added to two different control diets (See Table 1). All pigs were fed in groups of 4 or 6, of which half were females and half castrated males. The control diet A consisted of barley. As protein supplement each pig received daily 1 kg skimmed milk, 100 g soyabean meal and 50 g meat and bone meal. This diet was fortified with Ca, NaCl and vitamins A and D. In the control diet B approximately 20% of the energy in barley was replaced by boiled potatoes. The daily protein supplement was 1.5 kg

skimmed milk, 70 g soyabean meal and 35 g meat and bone meal
fortified with Ca, P, NaCl and vitamins A and D.

Experiment 1973 - 74

This comprised 8 replications of 4 treatments. Two
females and two castrated males were allotted to each pen.
The same feed mixture was used throughout the whole experimental period and consisted of 80% barley, 18% soyabean meal
and 2% minerals and vitamins. This diet contained 7 ppm Cu.
The mineral and vitamin mixture supplemented each kg of the
diet with 3 000 i.u. vitamin A and 1 000 i.u. vitamin D_3 and
in ppm: 100 zinc oxide, 125 ferrous sulphate, 125 manganese
sulphate, 5 cobalt sulphate, 1 potassium iodide, 5 riboflavin,
15 d-pantothenic acid, 20 vitamin E and 0.02 vitamin B_{12}. At
slaughter the whole liver from 8 pigs per treatment was
removed, deep frozen and ground. The copper content was
determined by atomic absorption spectrophotometry.

RESULTS

Experiment 1956 - 57

Table 1 summarises the results from the experiments of
1956 - 57.

When the diet was supplemented with 30 ppm Cu, there was
a 3% increase in daily gain and feed efficiency was improved
by 2%. Further supplement, up to 250 ppm Cu, did not increase
the response obtained by adding 30 ppm Cu to the diet. The
response to both daily gain and feed efficiency, however,
increased to approximately 7%, when 20% of the barley was
replaced by boiled potatoes (control diet B).

Investigations carried out by the Meat Research Institute
at Roskilde, did not show any difference due to the copper
supplementation with regard to pH, taste and colour of the
bacon.

-44-

TABLE 1

EXPERIMENT 1956 - 57

Control diet A			
Cu supplement, ppm	0	30	60
No. of pigs	32	32	32
Daily gain, g	590	606	600
FU per kg gain	3.39	3.31	3.34
Cu supplement, ppm	0	125	250
No. of pigs	8	8	8
Daily gain, g	595	607	602
FU per kg gain	3.06	3.02	3.02
Control diet B			
Cu supplement, ppm	0	60	
No. of pigs	12	12	
Daily gain, g	552	592	
FU per kg gain	3.42	3.14	

Experiment 1973 - 74

Table 2 shows that a supplement of 32 ppm Cu increased
daily gain and feed efficiency by 2%. This response was even
greater when the diet was supplemented with 125 ppm Cu in the
period 20 - 50 kg, while no response was observed from
50 - 90 kg. Thus, for the total period 20 - 90 kg, the
increase in daily gain and feed efficiency was 5%. No further
response was found when the supplement was increased from 125
to 200 ppm Cu. The results show too that in the last case the
Cu content of the liver increased from 4 to 16 ppm.

DISCUSSION

According to Landbrugsministeriets bekendtgørelse (1977)
the maximum copper level permitted in a complete feed mixture
for pigs in Denmark is 125 ppm. Supplements may be given as
Cu $(CH_3COO)_2.H_2O$, $CuCO_3.Cu(OH)_2.H_2O$, $CuCl_2.H_2O$, CuO and
$CuSO_4.5H_2O$. If the content of Cu exceeds 50 ppm it must be

declared by the manufacturer. Copper as sulphate, oxide and carbonate seems to have similar biological availability.

There is a general agreement that the requirement for copper is very small. Growing-finishing pigs need approximately 6 mg Cu/kg feed (Nielsen et al., 1979). As already mentioned copper may be used as a growth promoter although there is some controversy about the recommendations. In Danish experiments 250 ppm has been the highest level used (Table 1).

As will be seen the response in Table 2 is larger than in Table 1. Whether this is partly due to the higher protein content or sole use of vegetable protein in Experiment 1973 - 74 compared to Experiment 1956 - 57 cannot be concluded, however, from those experiments.

It is well known that interactions exist between different minerals like Cu, Zn, Fe and Ca. In the two series of experiments reported above the diet contained approximately 0.7% Ca and 0.6% P, but while the diet in Table 1 was not supplemented with Zn and Fe, the diet in Table 2 was supplemented with 80 ppm Zn and 25 ppm Fe, which may also have influenced the response obtained.

According to the Danish experiments in Table 2, 125 ppm Cu is sufficient from 20 to 50 kg and there is very little to gain by giving copper supplement from 50 to 90 kg.

It has been suggested that increased accumulation of liver copper may be responsible for the lesser response to copper during the finishing period. Recent experiments, however, have shown that a sulphide supplement may prevent high levels of liver copper whereas the growth promoting effect was still decreased.

TABLE 2

EXPERIMENT 1973 - 74

Cu supplement, ppm	O	32	125	200
No. of pigs	32	32	32	32
20 - 50 kg				
Daily gain, g	510	522	553[1]	545[1]
FU per kg gain	2.78	2.72	2.52[1]	2.60[1]
50 - 90 kg				
Daily gain, g	742	747	745	740
FU per kg gain	3.62	3.55	3.58	3.59
20 - 90 kg				
Daily gain, g	617	630	647[1]	641[1]
FU per kg gain	3.26	3.20	3.10[1]	3.16
Kg feed	219	214	208	212
Backfat, cm	2.43	2.37	2.43	2.50
Sidefat, cm	2.23	1.99	2.16	2.29
Percent lean meat	57.3	58.4	57.5	56.8
Iodine value in backfat	58.0	60.4	58.8	59.5
Weight of liver, g	1921	1691[1]	1797	1685[1]
Cu in liver, ppm	4.1	6.4	6.2	15.8[1]

1. Significantly different from the control group.

A large variation in liver copper was noticed between pigs
within the same group. Wegger and Ergün (1979) collected 200
liver samples from four slaughterhouses representing the whole
country. They found that the copper status was similar in all
4 parts of Denmark 7.9, 7.6, 7.6 and 7.4 mg/kg liver,
respectively. The mean was 7.6 ± 0.2 mg/kg liver and the
range was from 3 to 26. This may indicate a genetic or
environmental influence on the absorption and retention of
copper in the pig. While liver and kidney concentrations
reflect dietary copper, plasma Cu does not.

As pig producers often change to a lower protein diet
about 50 kg, it is possible to reduce or omit the copper in the

diet at this time and thus avoid too high accumulation of
copper in the pigs. If the diet is supplemented with 125 ppm
Cu in the period from 20 kg to 50 kg liveweight or from 20 to
90 kg, it amounts to administering approximately 10 and 28 g
Cu per pig in the two periods, respectively.

It is difficult to ascertain how much Cu is generally
present in pig diets. An investigation carried out by Kofoed
and Christensen (1979) shows the following contents of copper
in diets for bacon pigs:

    23 complete rations : 78 ppm Cu (SD = 48)
    79 protein supplements : 403 ppm Cu (SD = 158)

Some future aspects of the use of copper in agriculture
have recently been discussed by Dam Kofoed (1979) and he points
out the effects of using slurry from pigs fed copper supplemented diets, particularly when small areas of land are
available. In 21 samples of pig slurry he found 18 ppm Cu
(wet basis), while Madsen et al. (1976) found only 6 ppm Cu in
samples from an oxidation ditch. In the latter case pig diets
were supplemented with 30 ppm Cu. Finally it may be mentioned
that more than 90% of the ingested copper is excreted and this
may inhibit the anaerobic digestion process of pig waste.

REFERENCES

Braude, R., 1975. Copper as a performance promoter in pigs. Symp.
'Copper in Farming'. Copper Development Assoc., London. pp 79-97.

Clausen, Hj. et al., 1957. Kobbersulfat og zinkkarbonat. Bilag
Forsøgslab. efterårsmøde. p. 6.

Dam Kofoed, A., 1979. Kobber og dets anvendelse i landbruget. Ugeskrift
f. Jordbrug 124, 891-895.

Hansen, V., Sunesen, N. and Bresson, S., 1974. Kobbersulfat som
fodertilskud til slagterisvin. 416. Beretn. fra forsøgslab.,
København. 24 pp.

Kofoed, Th. and Christensen, S., 1979. Undersøgelser af indhold af
makro- og mikromineraler. 52. Beretn. Statens Foderstofkontrol,
Lyngby. pp. 55-57.

Landbrugsministeriets bekendtgørelse nr. 529 af 12. oktober 1977.
Bekendtgørelse om tilsaetningsstoffer til foderstoffer. p. 22.

Madsen, A., Nielsen, E.K., Huld. T., Scheel, B. and Pedersen, J., 1976.
Iltning af gylle II. Svinegyllens maengde, temperatur, sammensaetning
og biokemiske iltforbrug. 85. Meddelelse. Statens Husdyrbrugsforsøg.
4 pp.

Nielsen, H.E., Madsen, A. and Just, A., 1979. Danish recommendations of
minerals and vitamins for pigs. EAAP, Harrogate, England, July 1979.
5 pp.

Wegger, I. and Ergün, A., 1979. Mineralstofstatus hos svin bedømt ved
analyse af levervaev. Årsberetn. KVL's Institut f. Sterilitetsforsk.
pp. 135-145.

DISCUSSION

N.P. Lenis *(Netherlands)*

Do you think there is a setback in performance when you withdraw the copper at 50 kg liveweight?

A. Madsen *(Denmark)*

No, not at all, not according to our results.

N.P. Lenis

There are some reports which do indicate such a setback in performance.

A. Madsen

Yes I know; you can see this in Braude's review also. However, I was asked to present our Danish results and, as you saw, from 50 kg to 90 kg the figures were exactly the same for daily gain and for feed conversion.

N.P. Lenis

But what happens if you feed 20 ppm copper between 20 kg and 50 kg liveweight, and then stop feeding copper?

A. Madsen

I cannot answer this because we have not carried out such experiments.

PHYSIOLOGICAL ASPECTS OF COPPER IN PIG DIETS

J.B. Ludvigsen

National Institute of Animal Science,
Rolighedsvej 25, DK-1958 Copenhagen V, Denmark.

INTRODUCTION

Cu has attracted wide interest as a feed component for growing pigs, stimulating growth rate when given in amounts exceeding the concentration in the feed which is considered covering the optimal requirement for growth. Growth rate stimulation from 8 to 9.7% and increasing feed efficiency by 5.5 to 7.9% (Braude et al., 1962; Braude, 1967) have been demonstrated. Experiments using up to 207 ppm of Cu in the feed given as $CuSO_4$ conclude that the growth promoting effect of 132 ppm Cu was moderate, but better than that of 207 ppm in the first 8 weeks from 20 kg liveweight, whereas no effect was observed from 50 to 90 kg liveweight (Hansen et al., 1974). The variety of results from Cu supplement of the daily feed ration for pigs obviously depends on which Cu compounds have been used in the various experiments and how the macro- as well as the micro-mineral supply of the feed ration is balanced as Cu interaction with other minerals on the absorbative level is of unique importance in the intermediary metabolism of Cu.

ABSORPTION OF Cu

In common with other minerals, Cu is essential for the optimal function of intermediary processes of the growing as well as the adult pig. Cu is absorbed in ionic form from the stomach and the small intestine especially the upper small intestine. The rate of absorption is influenced by the amount of Cu, the chemical form of Cu ingested and the dietary level of other metal ions especially iron, zinc, molybdenum, cadmium, etc. because of a competitive level between Cu and Fe, Cu and Zn at the absorbative level. Calcium may be of

potential nutritional importance as high Ca levels in the feed
reduce Zn availability increasing the risk of Cu toxicity
(Underwood, 1977).

INTERMEDIARY METABOLISM OF Cu

Ionic Cu is very toxic and cannot exist in the body in
substantial amounts without toxic effects. Ionic Cu absorbed
has to be transferred to Cu-enzymes and other sites of
physiological actions, and the transport and distribution of
Cu to the sites of action is mediated through several typical
Cu-chelates, Cu-albumin and Cu-histidine, these transient
forms carrying Cu to its main store, the liver, where the
biosynthesis of ceruloplasmin takes place.

In healthy mammals 90% of the plasma Cu exists as
ceruloplasmin, a metalloid enzyme, often named ferrioxidase I,
an oxidase in iron utilisation promoting the rate of iron
saturation of transferrin in the plasma.

Cu can readily pass into erythrocytes. Around 60% of
total erythrocyte Cu is protein bound as erythrocuprein, an
active catalyser in oxygen metabolism. Part of the Cu in the
erythrocytes forms a second metallic component consisting of
two atoms of Zn and two atoms of Cu, another important
compound in intermediary oxidative pathways.

As mentioned, the intermediary metabolism of Cu is
closely connected with that of Fe. It is a well established
fact that Cu deficiency leads to anaemia in spite of adequate
Fe depots. It is suggested that 0.2 mg Cu/ml blood is the
minimum level at which haematopoiesis can take place in the
pig (Lahay et al., 1953). If such low levels are maintained
for any length of time anaemia is inevitable.

One of the important interactions of ceruloplasmin, the
key donor of Cu to the targets of metabolic action, is the
release of Fe from ferritin and the rate at which Fe ultimately

is transferred to Fe_3 (transferrin-complex) in haemoglobin synthesis. In Cu deficiency, low ceruloplasmin, Fe release from ferritin will be blocked followed by hypochromic and microcytic anaemia.

Generally, metalloproteins are of great importance in aerobic metabolism of cells as hydrogenation of oxygen is a prerequisite in the protection of the cells against two unavoidable by-products of oxygen reduction, superoxide and peroxide ions. These toxic intermediates are eliminated by a team of protective enzymes: superoxide dismutase (mainly Cu and Zn), Mn and Fe-enzymes, peroxidase and catalase. Further, cyrochrome oxidase, a Cu-haemoprotein, probably the single most crucial metalloenzyme of the mammalian cell in oxidative pathways, underlines the physiological importance of Cu.

The important interaction between Cu and Zn has received further support, as Zn-proteins have recently been demonstrated to be constituents of DNA and RNA polymerases, indicating the importance of the interactions of Cu, Fe and Zn in oxidative processes on the cellular level.

Cu DEFICIENCY AND FUNCTIONS

The manifestations of Cu deficiency vary with age and are more severe in the young growing pig than in adult pigs. Cu deficiency affects the function of the whole body, solely because of the vital importance of Cu in Fe metabolism. Manifestations of deficiency is a result of lack of Cu for all the metabolic processes involving Cu as a result of inadequate intake, depletion of body reserves or interaction with metabolic antagonists.

The interaction between Cu and Fe in the intermediary pathology of anaemia has been mentioned. Cu-deficient pigs have an impaired ability to absorb and mobilise Fe from the tissues for utilisation in the haemoglobin synthesis. The lifetime of erythrocytes is shorter in Cu-depleted pigs because

of an inadequate ceruloplasmin activity affecting the
mobilisation of Fe, particularly from the reticulo-endothelial
and/or the liver systems.

SKELETAL DISORDERS

Skeletal abnormalities occur in Cu-deficient pigs
(Teague et al., 1957). The histological changes are thinned
cortices, broadened epiphysial cartilage and low osteoblastic
activity (Follis et al., 1955).

The primary cause of the lesions in bone structures is
probably a reduction of the Cu-enzyme, amine oxidase, leading
to diminished stability and strength of bone collagen as the
result of an impaired cross-linkage of their polypeptide
chains.

CARDIOVASCULAR DISORDERS

Cardiac failure associated with hyperthropy of cardiac
musculature in the pig has been reported (Gubler et al., 1957).
The cytochrome oxidase activity of the heart muscle is reduced
affecting oxygen supply to the heart. Ceruloplasmin plays an
essential role in the recovery of the cytochrome oxidase
activity indicating that Cu is incorporated into cytochrome
oxidase only if it is presented to the cell as ceruloplasmin
(Broman, 1964).

The elastin content of the aorta of Cu-deficient pigs is
decreased (Weismann et al., 1963). The Cu-containing amino
oxidase enzyme is important for the incorporation of lysine in
the key cross-linkage groups in elastin. In Cu-deficient
animals amino oxidase activity in the aorta is reduced
resulting in less lysine being converted to desmosine, which
in turn results in less elasticity of the aorta.

Cu TOXICITY

Chronic as well as acute Cu poisoning may occur in the pig. Continued ingestion of Cu in excess of requirements leads to accumulation in the tissues, especially in the liver. Pigs are relatively resistant to excess amounts of Cu. Symptoms of Cu poisoning may occur at 250 ppm Cu in the feed, whereas pigs can tolerate 450 ppm and even more provided that the intake of iron and zinc is properly adjusted. The symptoms of Cu poisoning in pigs are a marked depression of appetite and growth rate, anaemia and jaundice.

The symptoms of toxicosis at the level of up to 450 ppm Cu in the feed can be eliminated by an additional supply of 150 ppm zinc plus 150 ppm iron.

At dietary levels of 750 ppm Cu, 500 ppm Zn plus 750 ppm Fe prevents symptoms of toxicosis. The protective effects of Fe and Zn against Cu toxicity can best be explained by the competition of the metal ions in the protein-binding processes at the absorbative stage and in the synthesis of metalloenzymes in the tissues.

Cu REQUIREMENTS OF THE PIG

As no differences were found in the performance of baby pigs fed a diet containing down to 6 ppm Cu, it was assumed that 6 ppm is adequate for growth of piglets (Ullrey et al., 1960). Later, as little as 4 ppm (ARC, 1967) was considered an optimal level of Cu in the feed of pigs up to 90 kg live-weight. Most feed compositions for pigs contain appreciably higher levels of Cu, e.g. grains contain 4 - 8 ppm. As Cu interacts metabolically with so many other elements, Zn and Fe as well as with Ca and Mo, it is very difficult to give maximum as well as minimum levels of Cu-requirements as the requirements will be determined by the amounts of Fe, Zn, Ca and other metal compounds in the feed. Thus, the Cu-requirement of the pig is within a wide range depending on the extent to which the interacting minerals are present or absent in the diet.

THE GROWTH PROMOTING EFFECT OF Cu

As mentioned earlier, increase in growth rate and a concomitant increase in feed conversion rate have been proven to take place in pigs when the diet is supplied with extra copper beyond the limits of physiological needs. It has been suggested that part of the improved performance might result from an anthelmintic effect of Cu. It has been suggested, too, that the effect might result from some interaction of Cu on the microflora of the intestinal tract, an effect that has failed to be proven.

Kirchgessner et al. (1976) found improved digestibility of protein in young pigs when adding extra Cu to the diet and demonstrated that cupric ions activite pepsin and raise peptic hydrolysis. Considering the way feeder pigs are fed, twice a day with an eating time of around 30 - 45 minutes per 24 hours which is far from the natural behavourial needs of the pig, an adequate pepsin secretion and peptic hydrolysis is a pre-requisite for the operation of the digestive tract, when feed intake is limited to two meals per day. A stimulation of this function may be part of the explanation of the improved performance.

However, part of the explanation of the improved weight gain and the decreased amount of feed per kg weight gain may theoretically be explained through a metabolic effect of high Cu supply reducing maintenance requirements, which also stimulates growth rate and reduces amount of feed per kg weight gain. An effect of extra Cu supplement on the overall protein and energy metabolism is still to be proven, but if it holds true excess of Cu may influence the activity particularly of the catabolic hormone systems, decreasing their activity on borderline physiological activities.

REFERENCES

Agricultural Research Council (UK), 1967. Nutr. Requir. Farm. Livest. No. 3.

Braude, R. et al., 1962. J. Agric. Sci. 58, 251.

Braude, R., 1967. World Rev. Anim. Prod. 3, 69.

Broman, L., 1964. Acta Soc. Med. Ups. 69, suppl 7.

Follis, R.H. et al., 1955. Bull. John Hopkins Hosp. 97, 405.

Gubler, C.J. et al., 1957. J. Biol. Chem. 244, 533.

Hansen, W. et al., 1974. 416. Rep. Nat. Inst. Anim. Sci., Copenhagen.

Kirchgessner, M. et al., 1976. Br. J. Nutr. 36, 15.

Lahay, M.E. et al., 1953. J. Clin. Invest. 32, 322.

Teague, H.S. et al., 1957. J. Nutr. 43, 389.

Ullrey, D.E. et al., 1960. J. Anim. Sci. 19, 1298.

Underwood, E.J., 1977. Trace Elements in Human and Animal Nutrition, Acad. Press.

Weismann, N. et al., 1963. J. Biol. Chem. 238, 3115.

DISCUSSION

K.L. Robinson (*UK*)

I have seen it claimed that a copper chelate - copper
methionate, is more effective as a feed additive for pigs than
the same level of copper as copper sulphate. Could Dr.
Ludvigsen comment on the question of absorption of copper
methionate from the gut and its stability in the gut, because
no evidence has been advanced on these subjects that I know of.

J. B. Ludvigsen (*Denmark*)

Probably there is some chelating effect in the gut, or
maybe in the epithelium of the small intestine, which would
explain a higher effect, but I don't really know. Do you
know, Dr. Braude?

R. Braude (*UK*)

I know of no firm evidence on this, I see no reason why
it should be so. One should remember that only 5% of dietary
copper is absorbed in any case.

K.L. Robinson

I wondered whether the stability of copper methionate in
the gut is such that it is absorbed as an intact chelate.

J.B. Ludvigsen

I don't know. What is known is the role of copper-
histidine chelate in transport in the blood and in the albumin,
but of course, methionine is in the albumin too.

I. Bremner (*UK*)

I think we are under a misapprehension if we keep saying
that only 5% of the copper in the diet is being absorbed. No
one has ever shown this. No one has ever made a direct
estimation of the true absorption of copper by the pig. Even
in rats it is very difficult to carry out these experiments.

What we do know is that 5% total retention may be taking place. There is a very efficient copper absorption process in many animals but the difficulty is that the liver is also highly efficient at excreting the copper again. A measurement of copper retention in the rat may give a figure of only about 5%, up to 95% being excreted in the faeces over one or two days after dosing. However, a measurement of true copper absorption could be found to be 50%.

R. Braude

How do you measure that?

I. Bremner

By whole body counting of the animals after dosing with copper-64. It is frequently said that copper absorption is an inefficient process. This is not so, in many cases it is efficient but the animal is well endowed with a homeostatic control mechanism which is located principally within the liver. The animal recognises if it has taken in too much copper and controls its total retention by pushing it out again, principally in the bile. So, if we are considering the effect of chelating agents on absorption, this may well be appreciable but it may be found subsequently that the liver can cope with this quite adequately by excreting the excess in the bile.

J.B. Ludvigsen

Yes, but if the organism is copper depleted and does not get an adequate copper supply, then you could expect a very high absorption of copper if you give the right component - maybe up to 90%.

R. Braude

What is the evidence for that statement? You are telling us the 90% can be absorbed.

J.B. Ludvigsen

Theoretically, yes.

M. Lamand (*France*)

I have not measured absorption in the pig but I have
measured apparent absorption in sheep and I agree that this is
rather poor. We find something like 10% maximum apparent
absorption.

A. Aumaitre (*France*)

Do you think there is any evidence for the effect of a
deficiency on the rate of absorption?

M. Lamand

In sheep, yes, digestibility is quite variable.
Obviously, different factors in the diet can modify the availability of copper. I did not measure true absorption by
labelled copper, but apparent absorption can be strongly
modified by factors in the diet.

THE EFFECT OF COPPER ON SOME PHYSIOLOGICAL AND BACTERIOLOGICAL CHARACTERISTICS IN DIFFERENT PARTS OF THE GASTRO-INTESTINAL TRACT OF GROWING PIGS.

N.P. Lenis

Institute for Livestock Feeding and Nutrition
Research (IVVO), Lelystad, The Netherlands.

ABSTRACT

From literature it is rather evident that the growth promoting effect of copper in pigs is due to actions on the microbial flora in the gastro-intestinal tract. These actions may result in modifications of the microflora in the small intestine, less production of growth depressing toxins and less degradation of essential nutrients. Two experiments were set up in order to get more information about the mode of action of copper sulphate. In a digestibility experiment some effects of Cu-addition were found in young pigs: somewhat higher digestibility coefficients for crude protein and lysine and less microbial transformation of cystine into methionine. In a slaughter experiment the effects of adding 155 ppm Cu to a diet on some physiological and bacteriological characteristics in different parts of the gut were investigated. There was some evidence that ammonia production, mainly in the large intestine, was inhibited by copper. Counting bacteria of the group of Enterobacteriaceae and Lactobacilli did not result in significant differences between treatments. Some intended analyses had to be cancelled because the samples got lost in a fire.

In conclusion, it can be said that copper addition might reduce degradation of essential nutrients (amino acids) in the small intestine, probably also resulting in less production of toxins; the Cu-response might be greater in young animals, emphasizing that environmental conditions influence the response.

INTRODUCTION

Because of its well-known growth promoting effect, copper addition to diets for growing-finishing pigs has been practised in the Netherlands for more than 20 years (at first 250 ppm, now up to 200 ppm total copper in the diet). The generally very positive response of growth rate and feed conversion ratio to addition of 125 - 250 mg Cu/kg diet, as found in many experiments, underlies this practice (Braude, 1967; Wallace, 1967; Meyer and Kröger, 1973; Borggreve, 1977).

While much research has been done in order to establish the quantitative effects of copper addition, proportionally little attention has been paid to the explanation of the mode of growth promotion by additional copper. Knowledge of the mode of action is very valuable, because it might provide a basis for developing alternative growth promoting substances. Finding a substitute for copper - its growth promoting effect seems to be additive to that of antibiotics - would be very attractive, especially in areas with an intensive pig production, because a prolonged use of high doses of copper may lead to an inacceptable Cu-accumulation in the soil.

SUGGESTED MECHANISMS OF GROWTH PROMOTION

A number of mechanisms have been proposed to explain the growth stimulatory effects of antimicrobial agents such as antibiotics and copper. Nearly all are based upon the concept that the growth response is due to actions on the microbial flora in the gastro-intestinal tract. Concerning copper we can even conclude that the action has to take place mainly in the proximal part of the small intestine. This is because in the distal part of the small intestine and especially in the large intestine, hydrogen sulphide is produced by bacterial fermentation of protein, which leads to the formation of inactive copper sulphide.

Some mechanisms frequently proposed are:

1. Copper might regulate the microbial flora in the small intestine and thus reduce the level of infectious disease;
2. Microbial production of growth depressing toxins might be reduced (which may lead to 'mechanism' 4);
3. Microbial destruction of essential nutrients in the gastro-intestinal tract might be reduced;
4. There might be an enhanced efficiency of absorption and utilisation of protein and/or energy.

Point 1.

The evidence that environmental conditions influence the response to antimicrobial agents (Visek, 1978) agrees with the remark made under 1. Several attempts have been made to establish quantitative changes in the microflora of the gastro-intestinal tract. However it appears to be very difficult to find reproducible results. Nevertheless Barber et al. (1961) and Hawbaker et al. (1961) found changes in the composition of the flora in faeces, brought about by copper supplementation. In the first part of the duodenum of pigs fed diets with 250 ppm copper, Dammers et al. (1959) found fewer *coli*-forms than in pigs fed diets without supplemental copper. In general they found a good correlation between the decrease of the coli-forms and the average daily gain. In contrast Kröger et al. (1977), sampling the end of the jejunum and the middle of the colon, could not find any effect on the number of *coli*-forms, but they found fewer *Clostridia perfringens* in the copper-supplemented groups.

Point 2.

There is evidence to suggest that ammonia is a factor in the antibiotic response (Visek, 1970; Easter and Baker, 1977). Antibiotics as well as copper - and other heavy metal-salts reduce desamination of amino acids and consequently ammonia production (Francois and Michel, 1956). This was also found by Dierick et al. (1978) (see also Henderickx et al. 1978) in *in vitro* studies with the ileum content of young pigs fed skim milk diets with additional antibiotics; in addition less decarboxylation products (amines) were found (especially the degradation of lysine was inhibited). Another toxic load for the animal, namely hydrogen sulphide, is probably relieved by copper feeding, not by inhibiting the liberation of it, but by reacting with copper-ions, producing the black, highly insoluble copper sulphide (Dammers et al., 1959; Braude, 1967).

Points 2. and 3.

Quantitative changes in the microflora as well as changes in the metabolism of intestinal microbes could underlie 'mechanisms' 2 and 3.

Point 3.

Effects of antimicrobial agents on the desamination of amino acids in the small intestine (see Point 2) probably mean that more essential nutrients are available for the animal. Vervaeke et al. (1978) calculated from *in vitro* experiments, carried out parallel to those of Dierick et al. (1978) (see Point 2), a protein and carbohydrate sparing effect, which correlated rather well with the growth promoting effect, observed *in vivo*. Just (1979) and Just et al. (1980) conclude from their own experiments and from those of Mason et al. (1976), that in cases of a small increase of the digestibility of amino acids, caused by antibiotic feeding, the main reason must be an increase in the digestibility of nitrogen and amino acids up to the terminal ileum and not a suppression of the microbial synthesis in the large intestine. Antibiotic feeding would rather inhibit the disappearance of nitrogen-containing substances in the large intestine.

Points 3. and 4.

Kirchgessner and Giessler (1961) found an increase in the nitrogen digestibility and retention in young pigs on diets with 250 ppm copper. Dammers et al. (1959) using two animals per diet, also observed higher digestibility

coefficients for crude protein, crude fat and crude fibre on
a diet with 250 ppm copper. However, Castell and Bowland
(1968), using diets with 250 ppm copper could only observe a
better nitrogen digestibility at a liveweight of the pigs of
25 kg. Farries and Angelowa (1967) could not find any effect
of copper on nitrogen digestibility or retention.

Point 4.

As observed frequently in antibiotic-fed chickens, swine
and rats (Visek, 1978) Dammers et al. (1959), feeding diets
with 250 ppm copper showed a tendency to reduce the weight of
the small intestine, probably caused by a decrease in the
thickness of the small intestinal wall.

OWN EXPERIMENTS

In order to obtain more information about the mode of
action of copper sulphate two experiments were set up,
examining the effect of copper on the digestibility of crude
protein and amino acids and the effect on some physiological
and bacteriological characteristics in different parts of the
gastro-intestinal tract. The latter was done with young pigs,
because the response is believed to be better in young growing
pigs (Meyer and Kröger, 1973).

MATERIALS AND METHODS

Experiment 1.

Digestibility trials were carried out with three diets
for growing finishing pigs, containing 8, 106 and 216 ppm
copper (as $CuSO_4.5 H_2O$) respectively. The diets (mainly
composed of cereals and soyabeanmeal) contained 8.5 MJ/kg net
energy for pigs, 14.0% crude protein and 0.62% lysine. Each
diet was tested three times with three castrated male pigs at
average liveweights of 36, 67 and 99 kg. The same animals
received the same diet during the whole experiment. Faeces
were collected quantitatively twice a day during a 10-day period.

Formalin was added to the collected faeces for preservation.
after which they were stored at 2 - 4°C. All analyses were
carried out on the fresh material. For amino acid analysis an
automatic analyser was used.

Experiment 2.

In a slaughter experiment two high quality diets, one
with 15 ppm copper (only 2.5 ppm added) and one with 170 ppm
copper (155 ppm added) were fed to 24 piglets (12 boars and
12 sows) in the liveweight interval ca. 16 - ca. 50 kg. To
both diets 80 ppm Fe and 60 ppm Zn were added; apart from
copper no other antibacterial agent was used. The diets included a.o. 15% skimmed milk powder (not denaturated) and
contained 9.2 MJ/kg net energy for pigs, 18.3% crude protein
and 1.08% lysine. The animals of both treatments were housed
in different pens without straw and were fed individually
according to a restricted scale. Differences in feed intake
between treatments did not occur, because two animals of the
same sex, one from the additional copper-group and one from the
control group (littermates), always received the same quantity
of feed.

Pigs were killed at the beginning of the experiment (4
animals) and at specific intervals during the experiment (12
pairs of animals) by injecting triotal intravenously and
samples of gut contents were prepared as quickly as possible.
The gastro-intestinal tract was cleaned and weighed. Killing
occurred on day 28, 37, 49, 50 and 51 after starting the
experiment, 2 - 4 hours after feeding. The interval between
killing two pigs of one pair was always approximately half an
hour. Killing of pigs within a pair was arranged according to
an alternate scheme. In the gastro-intestinal tract, samples
were taken from: 1) the stomach; 2) the duodenum, just beyond
the entrance of the pancreatic duct; 3) the middle of the
jejunum; 4) the distal end of the ileum (ca. 50 cm); 5) the
caecum; 6) the large intestine, between 50 - 100 cm distal of
the caecum; 7) the distal 50 cm of the large intestine; 8) the
rectum.

After opening the abdominal cavity the parts of the
gastro-intestinal tract concerned were tied up, the samples
were taken immediately and cooled in ice. After measuring the
pH the samples were prepared for determining dry matter, crude
protein and ammonia, the latter according to the Berthelot-
method.   In samples 2, 4 and 6 of 4 pairs of boars and 4 pairs
of sows, killed on day 49 - 51, some bacteriological counts were
made by the Department of Bacteriology of the Faculty of
Veterinary Medicine in Utrecht. The remainder of each sample
was frozen in order to carry out some other analyses later
(amines according to a new method, developed by Dierick et al.
(1976) and volatile fatty acids. Concerning bacteriology,
counts were made of the group of *Enterobacteriaceae* according
to the violet red bile dextrose agar method, formulated by
Mossel et al. (1962) and of the group of *Lactobacilli* according
to the Rogosa agar method, described by Rogosa et al. (1951)
and modified by Sharpe (1960).

RESULTS

Experiment 1.

Figures for the apparent digestibility of crude protein
and the amino acids lysine, threonine, methionine and cystine -
often limiting amino acids - are presented in Table 1. The
digestibility coefficients of the other amino acids, dry matter,
crude fat, crude fibre and nitrogen free extract are not shown;
no real differences between treatments were found.

Experiment 2.

During the first month of the experiment diarrhoea was
more frequently observed in the control group than in the
copper-supplemented group. However addition of copper did not
result in improvement of daily gain or feed conversion ratio.
The performance data of 4 pairs of boars and 4 pairs of sows.
killed on day 49 - 51, are presented in Table 2.

TABLE 1

APPARENT DIGESTIBILITIES OF CRUDE PROTEIN AND SOME AMINO ACIDS IN DIETS I (8 ppm Cu), II (106 ppm Cu) AND III (216 ppm Cu) IN EXPERIMENT 1. (IN %)

Live-weight (kg)	Diet	Crude protein		Lysine		Threonine		Methionine		Cystine	
		mean	s.d.	mean	s.d.	mean	s.d.	mean	s.d.	mean	s.d.
36	I	73.5	2.5	64.5	4.8	66.2	2.6	55.0	12.0	80.9	2.5
	II	*		69.5	1.8	68.8	1.5	64.5	4.8	81.4	2.2
	III	76.1	1.5	69.4	2.8	68.4	1.4	63.7	2.8	73.0	0.5
67	I	79.3	1.0	77.0	0.9	76.1	0.6	68.5	4.6	78.1	0.8
	II	79.4	0.5	74.4	1.2	74.1	1.1	70.0[1]		74.2	2.1
	III	78.4	1.2	75.0	1.3	74.8	0.9	73.4	5.0	74.2	1.9
99	I	80.9	0.1	77.8[2]	2.9	74.9[2]	4.3	75.8[2]	0.5	78.3[2]	5.4
	II	79.5	0.7	76.5	0.6	75.3	0.6	78.3	1.9	78.6	2.5
	III	81.1	0.5	79.1	0.7	76.8	0.1	79.3	1.4	77.0	0.9

* The samples got lost

1) Only one animal

2) Average of only two animals

TABLE 2

DAILY GAIN AND FEED CONVERSION RATIO OF 4 PAIRS OF BOARS AND 4 PAIRS OF SOWS, KILLED ON DAY 49 - 51

| Sex | + Cu | | - Cu | |
	daily gain (g/day)	feed conv. ratio	daily gain (g/day)	feed conv. ratio
Boars	633	2.04	627	2.03
Sows	634	2.04	659	1.98

Data concerning pH-measurements and ammonia- and crude protein analyses in the digesta of several places of the gastro-intestinal tract are presented in Table 3. Those from the four animals killed at the beginning of the experiment have been omitted here because the effect of copper addition was invest-igated by comparing within pairs. With relation to the weight of the emptied gastro-intestinal tract no differences between treatments were found; these figures have also been omitted.

Concerning bacteriology, the results of counting the *Enterobacteriaceae* and the *Lactobacilli* are shown in Table 4. Unfortunately, analysing the frozen samples for amines and volatile fatty acids had to be cancelled, because all the samples got lost in a heavy fire in the Institute's laboratory in January 1979.

DISCUSSION

Experiment 1.

From the figures in Table 1 it may be concluded that copper addition resulted in somewhat higher digestibility coefficients of crude protein and amino acids in the young pigs (36 kg). The improved lysine digestibility would be in agree-ment with the observations made by Dierick et al. (1978) that especially the degradation of lysine to cadaverine was inhibited by virginiamycine and spiramycine. However the standard deviations were rather high and in the older pigs no

-69-

TABLE 3

pH-MEASUREMENTS AND NH_3 - (mg/100 ml) AND CRUDE PROTEIN (% IN DRY MATTER) ANALYSES IN GUT CONTENTS OF SIX PAIRS OF BOARS AND SIX PAIRS OF SOWS, KILLED ON DAY 28 - 51

Sex	C[1]	T[2]	Sampling place *															
			1		2		3		4		5		6		7		8	
			mean	s.d.	mean	s.d.	mean	s.d.	mean	s.d.	mean	s.d.	mean	s.d.	mean	s.d.	mean	s.d.
Boars	pH	+Cu	4.01	0.41	5.23	0.41	6.46	0.32	6.55	0.34	5.39	0.19	5.49	0.22	5.88	0.41	6.19	0.36
		-Cu	3.78	0.29	5.10	0.43	6.29	0.28	6.57	0.24	5.32	0.11	5.48	0.08	6.04	0.35	6.28	0.35
	NH_3	+Cu	26.3	8.9	16.7	3.7	17.8	5.4	16.8	5.8	12.3	7.9	24.7	4.9	27.8	7.8	37.9	5.7
		-Cu	27.8	7.7	21.9	6.7	19.8	6.1	18.1	0.6	13.9	8.3	27.7	12.6	50.3	17.1	51.9	5.9
	CP	+Cu	18.1	2.9	31.0	9.4	27.1	7.7	24.1	7.8	23.5	5.0	24.9	3.2	24.5	3.4	26.2	3.6
		-Cu	16.6	2.0	27.1	4.9	27.5	7.7	22.1	4.6	22.1	2.9	26.1	2.8	25.8	2.3	26.2	3.0
Sows	pH	+Cu	4.11	0.50	5.44	0.16	6.52	0.27	6.58	0.27	5.45	0.18	5.53	0.08	5.79	0.35	6.40	0.11
		-Cu	3.85	0.41	5.16	0.38	6.38	0.23	6.58	0.23	5.38	0.09	5.53	0.10	5.90	0.15	6.37	0.23
	NH_3	+Cu	29.3	7.7	22.1	9.0	23.8	4.6	12.8	7.0	13.0	6.9	27.8	5.2	33.5	5.5	41.0	15.7
		-Cu	33.8	11.7	24.8	7.0	34.0	7.4	18.5	4.2	13.7	3.3	29.5	10.3	45.6	9.3	49.3	12.9
	CP	+Cu	21.1	7.7	29.2	2.0	26.8	2.8	20.2	2.3	27.9	8.6	24.5	2.0	23.6	1.3	27.2	4.3
		-Cu	18.8	3.0	34.2	8.1	28.4	4.8	19.8	2.8	22.2	1.9	24.0	1.1	24.0	1.5	25.1	1.7

* For explanation see 'Material and methods'.
1) C = Criterion; 2) T = treatment.

TABLE 4

DILUTION PROCEDURE AND COUNTS OF 2 GROUPS OF BACTERIA IN GUT CONTENTS OF
4 PAIRS OF BOARS AND 4 PAIRS OF SOWS, KILLED ON DAY 49 - 51

Animals	Sampling place*	Dilution	*Enterobacteriaceae* + Cu Number of colonies per ml	− Cu Number of colonies per ml	Dilution	*Lactobacilli* + Cu Number of colonies per 0.1ml	− Cu Number of colonies per 0.1ml
Boars 1 + 2	2	10^{-3}	3	1	10^{-5}	20	66
	4	10^{-4}	105	10	10^{-6}	90	36
	6	10^{-5}	562	4	10^{-6}	75	178
Boars 3 + 4	2	10^{-3}	14	5	10^{-6}	54	22
	4	10^{-5}	95	2	10^{-6}	27	105
	6	10^{-5}	106	1	10^{-6}	53	400
Boars 5 + 6	2	10^{-3}	27	9	10^{-4}	260	265
	4	10^{-5}	100	130	10^{-6}	28	9
	6	10^{-5}	320	59	10^{-6}	30	35
Boars 7 + 8	2	10^{-3}	-	32	10^{-5}	22	57
	4	10^{-5}	123	95	10^{-5}	250	149
	6	10^{-5}	195	80	10^{-6}	62	71
Sows 1 + 2	2	10^{-3}	104	27	10^{-4}	625	99
	4	10^{-5}	43	500	10^{-6}	42	25
	6	10^{-5}	56	168	10^{-6}	98	301
Sows 3 + 4	2	10^{-3}	440	2	10^{-4}	197	225
	4	10^{-4}	44	64	10^{-5}	15	15
	6	10^{-4}	12	140	10^{-6}	123	30
Sows 5 + 6	2	10^{-3}	59	3	10^{-6}	18	3
	4	10^{-6}	155	1	10^{-6}	55	43
	6	10^{-6}	94	2	10^{-6}	200	38
Sows 7 + 8	2	10^{-4}	-	93	10^{-5}	4	32
	4	10^{-5}	4	84	10^{-5}	5	238
	6	10^{-5}	15	152	10^{-6}	154	179

* For explanation see 'Material and methods'

improvement in digestibility could be observed. The latter
would be in agreement with the observations made by
Kirchgessner and Giessler (1961) and Castell and Bowland (1968).
Concerning the sulphur-containing amino acids it is striking,
that the cystine digestibility in the high-copper group at
36 kg liveweight was significantly lower than in the other
groups. At the same time methionine digestibility increased
from diet I to diet III especially in the young pigs. This
points to less bacterial activity in the large intestine of
the pigs, fed the high copper diet: less cystine was trans-
formed into methionine in the large intestine (Lenis, 1980).
Also at 67 kg liveweight the same tendency was found. The
greater effect, found in young pigs, suggests that environmental conditions influence the response. Another striking
phenomenon is the big gap between digestibility coefficients,
determined at 36 and 99 kg liveweight, especially in the non-
supplemented group.

Although some effects of Cu-addition on digestibility
were found, it is questionable whether the faeces-analysis-
method is suitable for investigating effects of growth-
promoting agents, because the main effect is probably located
in the small intestine; microbial transformations of nitrogen-
containing substances in the large intestine - in general not
relevant to the animal - might be very confusing (Just, 1979;
Just et al., 1980; Lenis, 1980). So ileal digestibility for
these kind of studies is probably preferable.

Experiment 2.

Although no growth promoting effect was found, possible
effects of addition of copper on the microflora might have
taken place. Concerning pH-measurements no great differences
were observed between treatments (table 3). Nevertheless pH-
values in the stomach and the proximal part of the small
intestine were somewhat higher in pigs fed the Cu-supplemented
diet. The pH-values, found in the caecum, large intestine and
rectum were somewhat lower than found in other experiments
(Schulze and Bathke, 1977).

From Table 3 can be concluded that copper addition inhibited ammonia production in the large intestine. In the small intestine only a tendency to a small inhibiting effect was found. So, due to a rather great variation definitive conclusions about the effect of copper on production of ammonia, a potential toxin, in the small intestine - where the main action probably takes place - cannot be drawn. Nevertheless, the data confirm the indications found in other experiments, that to a certain extent copper reduces microbial production of growth depressing toxins and microbial destruction of amino acids in the small intestine.

Concerning the counts of bacteria, no real differences between treatments were found for the group of *Enterobacteriaceae* and the group of *Lactobacilli*. The variation was rather high. The results (number of bacteria/g intestinal content) are in good agreement with other studies (Schulze and Bathke, 1977). From this study and other studies it is becoming evident that bacteriological work is not very suitable for research into the mode of action of growth promoting agents.

CONCLUSION

In this study copper addition resulted in a higher digestibility of crude protein and amino acids in young pigs. Also a tendency to inhibition of ammonia production was found. So from this study and from other studies in literature it may be concluded that copper addition might reduce degradation of essential nutrients (amino acids) in the small intestine, probably also resulting in less production of toxins. The Cu-response might be greater in young animals, emphasizing that environmental conditions influence the response.

ACKNOWLEDGEMENT

Appreciation is expressed to the Commissie Hinder-preventie (Commission Nuisance Prevention Livestock-operations) of the Commodity Board of Feedstuffs in the Hague for giving financial support for this study.

REFERENCES

Barber, R.S., Braude, R. and Mitchell, K.G., 1961. Copper sulphate and
copper sulphide (CuS) as supplements for growing pigs. Brit. J.
Nutr., 15, 189.

Braude, R., 1967. Copper as a stimulant in pig feeding (*cuprum pro
pecunia*). World Rev. Anim. Prod., 3, 69.

Borggreve, G.J., 1977. Het groeibevorderende effect van extra koper voor
varkens. Literatuuroverzicht en samenvattend verslag van zeven
voederproeven op De Schothorst. C.L.O.-Instituut voor de Veevoeding
"De Schothorst", Hoogland.

Castell, A.G. and Bowland, J.P., 1968. Supplemental copper for swine:
growth, digestibility and carcass measurements. Can. J. Anim. Sci.,
48, 403.

Dammers, J., Stolk, K., van der Grift, J. and Frens, A.M., 1959. Het toe-
voegen van kopersulfaat aan rantsoenen voor mestvarkens. Verslagen
van Landbouwkundige Onderzoekingen no. 65. 12. Pudoc, Wageningen

Dierick, N.A., Vervaeke, I.J., Decuypere, J.A. and Henderickx, H.K., 1976.
Determination of biogenic amines in intestinal contents by ion-
exchange chromatography. J. Chromatogr., 129, 403.

Dierick, N.A., Vervaeke, I.J., Decuypere, J.A. and Henderickx, H.K., 1978.
Aminozuurdegradatie onder invloed van de darmflora van de big: *in
vitro* benadering. Personal communication.

Easter, R.A. and Baker, D.H., 1977. Arginine and its relationship to the
antibiotic growth response in swine. J. Anim. Sci.,45, 108.

Farries, E. and Angelowa, L., 1967. Ein Beitrag zur Stickstoffverwertung
nach Kupfersulfatzulage beim Schwein. Landwirtsch. Forschung,
20, 137.

Francois, A. and Michel, M., 1956. Le rôle des antibiotiques dans la
croissance. Ann. de la Nutr. et de l'Aliment., 10, 94.

Hawbaker, J.A., Speer, V.C., Hays, V.W. and Catron, D.V., 1961. Effect of
copper sulphate and other chemotherapeutics in growing swine rations.
J. Anim. Sci., 20, 163.

Henderickx, H.K., Vervaeke, I.J., Decuypere, J.A. and Dierick, N.A., 1978.
Die Wirkung von Wachstumsfördern auf die biochemischen Leistungen
von Darmbakterien beim Schwein. Z. Tierphysiol., Tierernährg. u.
Futtermittelkde.,40, 118.

Just, A., 1979. Ileal digestibility of protein: applied aspects. In
  'Current concepts of digestion and absorption in pigs'. (A.G. Low
  and I.G. Partridge, editors) Reading: National Institute for Research
  in Dairying, Technical.Bulletin $\underline{3}$, 66

Just, A., Sauer, W.C., Bech-Andersen, S., Jørgensen, H.H. and Eggum, B.O.,
  1980. The influence of the hind gut microflora on the digestibility
  of protein and amino acids in growing pigs elucidated by addition of
  antibiotics to different fractions of barley. Z. Tierphysiol.,
  Tierernährg. u. Futtermittelkde., $\underline{43}$, 83.

Kirchgessner, M. and Giessler, H., 1961. Der Einfluss einer $CuSO_4$-Zulage
  auf den N-Ansatz wachsender Schweine. Z. Tierphysiol. Tierernährg.
  u. Futtermittelkde. $\underline{16}$, 297.

Kröger, H., Feder, H., Plischke, R. and Amtsberg, G., 1977. Untersuchungen
  über den Einfluss verschiedener Cu-Zusätze (Kupfersulfat, Kupfer-II-
  Oxid und elementares Kupfer) auf die Mast-und Schlachtleistungen beim
  Schwein. Züchtungskunde, $\underline{49}$, 213.

Lenis, N.P., 1980. De aminozuurverteerbaarheid bij het varken en de rol
  van de bacterieflora in de dikke darm en de bijdrage van het endogene
  eiwit. Rapport no. 133. Institute for Livestock Feeding and
  Nutrition Research (I.V.V.O.), Lelystad.

Mason, V.C., Just, A. and Bech-Andersen, S., 1976. Bacterial activity in
  the hind-gut of pigs. 2) Its influence on the apparent digestibility
  of nitrogen and amino acids. Z. Tierphysiol., Tierernährg. u.
  Futtermittelkde, $\underline{36}$, 310.

Meyer, H. and Kröger, H., 1973. Kupferfütterung beim Schwein. Übers.
  Tierernährg. $\underline{1}$, 9.

Mossel, D.A.A., Mengerink, W.H.J. and Scholts, H.H., 1962. Use of a
  modified MacConkey Agar medium for the selective growth and enumeration of *Enterobacteriaceae*. J. Bact., $\underline{84}$, 381.

Rogosa, M., Mitchell, J.A. and Wiseman, R.F., 1951. A selective Medium
  for the Isolation of Oral and Fecal *Lactobacilli*. J. Bact., $\underline{62}$, 132.

Sharpe, M. Elizabeth, 1960. Selective Media for the Isolation and
  Enumeration of *Lactobacilli*. Lab. Practise $\underline{9}$, 223.

Schulze, F. and Bathke, W., 1977. Zur quantitativen Zusammensetzung der
  Magen-Darm-Flora beim Läuferschwein. Arch. exper. Vet. med., 31, 161.

Vervaeke, I.J., Decuypere, J.A., Dierick, N.A. and Henderickx, H.K., 1978.
Kwantitatieve benadering van het groeibevorderingspercentage door
antibiotica bij het varken, steunend op *in vivo* en *in vitro*
gegevens. Personal communication.

Visek, W.J., 1970. Urease: its significance in birds and mammals. Agr.
Sci. Rev., 1st qtr. 1970: 9.

Visek, W.J., 1978. The mode of growth promotion by antibiotics. J. Anim.
Sci., 46, 1447.

Wallace, H.D., 1967. High level copper in swine feeding. Internat. Copper
Research Association, New York.

DISCUSSION

J.B. Ludvigsen *(Denmark)*

Dr. Lenis, in your bacteriological study you say '......
only coliform bacteria'. Is that a mixture of non-haemolytic
and haemolytic *coli* ?

N.P. Lenis *(Netherlands)*

Yes, it is the total group of *Enterobacteriacea* . I am
aware that it was not a very detailed study but it was intended
to be a preliminary study and it is very difficult to go into
the bacteriological aspects in greater depth.

J.B. Ludvigsen

This is true, but if you had stated the amount of
haemolytic *coli* and maybe the strains, it would be easier to
evaluate the possible toxic effect.

R. Braude *(UK)*

On the bacteriological side, I fully agree with the
bacteriologists that there is very little to be gained from
this sort of analysis. However, as far as the effect on
digestibility of amino acids is concerned, we would query that
approach nowadays. It is completely meaningless; the faecal
digestibility of amino acids is a meaningless term.

N.B. Lenis

Yes. I made this point in my paper; I have said that
ileal digestibility is preferable for this kind of study.

GENERAL DISCUSSION

J.K.R. Gasser *(UK)*

I have been listening to the papers this morning with a great deal of interest, and particularly the description of the experiments to which I would apply the criticism often levelled at field experiments on plants, that they are empirical in nature. Although one can obtain general benefits and increases, and a lot of our recommendations have to be based on such experiments because we have no better method at the present time, nevertheless empirical experiments do not provide us with information on underlying mechanisms. For example, experiments testing the effectiveness of copper as a feed additive do not provide information on why one pig in six does not respond to copper.

Speakers then considered some more detailed aspects of copper metabolsim and why and how it may affect the growth performance of pigs. In this general discussion I believe we should try to think constructively about the type of work we would like to see done in the future in order to fill some of the very obvious gaps. For example, there is a very timely aspect in genotype and breed effects on response to copper.

B.C. Cooke *(UK)*

May I comment on the breed aspect. Very little difference was found by workers in the United States, Canada and the UK, where the breeds are significantly different, suggesting that there is no breed genotype effect on the response.

J.K.R. Gasser

This means that we need to look for some other factor which is affecting the performance of individual pigs.

I. Bremner *(UK)*

It is well established in other species, such as the

sheep, that there are major differences between breeds in susceptibility to copper deficiency, and also to copper poisoning.
It would be surprising if the pig is different in this regard.

J.K.R. Gasser

Could this be a difference between the ruminant and the
non-ruminant?

I. Bremner

No, I don't think so. I can cite examples from laboratory
mice and rats where different strains seem to have major differences in basic aspects of copper metabolism.

R. Braude (UK)

Dr. Bremner is postulating that there may be a breed
effect. The effect of feeding high levels of copper has been
studied on all the existing breeds of pigs in the world, with
the same variation of response. To postulate that there is a
genetic factor in the response to the level of copper that we
are recommending to feed is, I think, not likely.

I. Bremner

The comments I made were in relation to basic aspects of
copper metabolsim; they were not concerned with growth promoting effect, which is probably a different phenomenon.

J.H. Voorburg (Netherlands)

My problem is not the 5% of copper which is digested by
the pig but rather the 95% in the slurry. Is it possible to re-
feed this copper to the pig with the same effect as copper sulphate?

A. Aumaitre (France)

I would like to comment on the recycling of pig waste.
For example, in considering the results from Harmon and his co-
workers in Illinois, only the top layer of the slurry in the

oxidation ditch was used for feeding the animals, which I
believe was from pigs not receiving extra dietary copper. If
you look at the content of copper in faeces and urine of animals receiving 250 ppm copper, you will find values from about
800 to 1 300 ppm (Table 9 of my paper). Therefore, in re-
feeding there is a danger of going beyond the toxic level which
would be around 300 - 500 ppm. One must also be cautious in
re-feeding pig wastes because of the high content of other
minerals such as phosphorus and calcium.

J.H. Voorburg

But the question I was asking was this: if you re-feed
in an appropriate way, will the copper in the waste have the
same effect as the copper sulphate which you add to the ration?

J.K.R. Gasser

After discussion the conclusion was reached that this has
not yet been tried and so there is no definitive answer.

A. Madsen (Denmark)

As Dr. Braude pointed out, pigs vary in their response to
added copper from -15 to +30%. I would like to know whether
the copper has any influence on the quality of the feed mixture.
Sometimes the quality of the feed mixture deteriorates with
little daily gain by the pigs and lessened growth rate. Could
copper have an anti-fungal effect so that when you supplement
the feed with copper, you are maintaining the quality? Have
there been any experiments in which copper has been added
immediately prior to feeding, thus establishing that the copper
has a direct effect rather than a general effect of maintaining
the quality of the feed during storage?

R. Braude

The theory of copper having a possible anti-fungal effect
has been postulated many times and it has been compared with
other important anti-fungal agents. Copper has been found to

have an effect where anti-fungal agents have not. That is the
best answer we can give to that question.

D.B.R. Poole *(Ireland)*

Some years ago, a colleague of mine observed a very
strong effect of copper in feed as a pro-oxidant, increasing
the level of rancidity of the fat. He associated this with
degradation of unprotected vitamins in the feed.

K.L. Robinson *(UK)*

Dr. Lenis raised several bacteriological questions in
his paper. Research in this area is being done, particularly
in Germany, which provides some information on the points
raised by Dr. Lenis on the possible role of copper in stabil-
ising gut flora, therefore making it less easy for disease
promoting organisms to establish themselves in the presence of
copper. This, of course, raises questions of the continuation
of added copper in the diet, and the amount added. There is a
very difficult problem whether copper has a role in inducing
resistances to antibiotics in gut bacteria. So, I hope that
we will hear more about microbiological aspects of copper in
pig diets.

N.P. Lenis *(Netherlands)*

Environmental conditions are clearly very important for
the response to added copper, particularly in young pigs. We
have to conclude that copper acts mainly at the level of the
gastro-intestinal tract and the effects are metabolic.

R. Braude

Unfortunately, we are lacking evidence on germ-free pigs
and it is very difficult to get such evidence.

SESSION II

AGRONOMY

A) SOIL PLANT RELATIONSHIPS

Chairman: J.H. Voorburg

BEHAVIOUR OF COPPER IN SOILS:
ADSORPTION AND COMPLEXATION REACTIONS

L. Kiekens and A. Cottenie

Laboratorium voor Analytische en Agrochemie, Faculteit van
de Landbouwwetenschappen, Rijksuniversiteit Gent,
Gent, Belgium.

ABSTRACT

Adsorption of Cu by soils was fitted to the Langmuir adsorption equation. The calculated Langmuir parameters b (adsorption maximum) and k (bonding energy coefficient) were closely related to soil characteristics. The presence of 0.01 n $CaCl_2$ reduced Cu adsorption, indicating that Ca^{2+} competes effectively with Cu^{2+} for the adsorption sites. The influence of pH on Cu adsorption was shown by increasing adsorption at higher pH values.

An experimental technique is described, permitting the estimation of the distribution of added Cu in soils as a function of pH. At the actual soil pH only a very small fraction of added Cu remained in solution. A relatively small fraction of retained Cu was exchangeable with 1 n NH_4OAc pH $\doteq$ 7; while some 15 - 25% was extractable with 0.2 n NaOH, which is a measure for Cu bound to organic substances. Soil organic matter may be considered as an important regulator of Cu mobility in soils. Humic acids mainly form insoluble complexes in acid medium and soluble ones in alkaline conditions. However, fulvic acid complexes are less stable than the corresponding humates.

INTRODUCTION

Availability of trace elements in soils largely depends on their interactions with various constituents such as organic matter, mineral colloidal particles (clay), carbonates and amorphous oxides of Fe, Al, Mn. The type of interaction which will be the most critical in a given system depends on several parameters, namely:
- the concentration of ions in the soil solution,
- type and abundance of adsorption sites associated with the solid phase,
- concentration of ligands, able to form organomineral complexes,
- pH and redox potential.

These parameters mainly control the equilibrium reactions in which trace elements are involved, namely, precipitation and dissolution, complexation and decomplexation, adsorption and desorption. The equilibrium constants of these reactions govern the distribution of trace elements between different forms. Besides the amounts retained in the solid phase, trace elements may be present as free ions in the soil solution, as soluble organo-mineral complexes and adsorbed at soil colloids. Equilibrium displacements, resulting in a transfer from one form to another may occur as a consequence of changing physical and chemical conditions.

The purpose of this paper is to show the influence of organic matter and mineral soil colloids on the behaviour of Cu in soils.

ADSORPTION OF Cu BY SOILS

Literature review

Generally soil colloids act as a negative ion exchanger and hence may remove cations from the soil solution through ion-exchange reactions. Thus soil colloids may be considered as a buffering system to regulate the concentration of cations

in the soil solution. Adsorption of Cu by soils has received
increasing attention in recent years, as may be concluded from
the multitude of publications dealing with this topic.

Several investigators used the Langmuir equation to
describe adsorption of Cu by soils (McLaren and Crawford, 1973;
Udo et al., 1970; Shuman, 1975; Levi-Minzi et al., 1976;
Petruzelli et al., 1978) and showed that Cu adsorption conformed to the Langmuir equation.

Bower and Truog (1941), Elgabaly and Jenny (1943),
DeMumbrum and Jackson (1956) found that soil colloids,
especially at neutral and alkaline pH, retained Cu in excess
of the cation exchange capacity. An explanation for the excess
retention may be the formation of $CuOH^+$, $CuCl^+$ and other
complexes, or precipitation of Cu compounds. In calcareous
soils this phenomenon may be explained by the formation of Cu-
carbonates (Misra and Tiwari, 1966). According to Bingham
et al. (1964) excess retention is mainly due to precipitation
of $Cu(OH)_2$. At pH values at which precipitation is very low,
these authors found the same analogous retention characteristics as for other divalent cations.

Several authors found that Cu may be specifically
adsorbed by soil colloids. According to McLaren and Crawford
(1973) organic matter and free Mn oxide content of the soil
seem to be the most important characteristics contributing to
the specific adsorption of Cu.

Cavallaro and McBride (1978) evaluated adsorption,
precipitation and complexation as possible parameters for controlling the concentrations of Cu and Cd in the soil solution.
It was shown that Cu was more strongly retained than Cd, due
to specific complexation reactions with humic and fulvic acids.
Analogous conclusions can be drawn from the studies of
Petruzelli and Guidi (1976), Davies et al. (1969), Plessis and
Burger (1971), Guidi et al. (1972), Schlichting (1955), Szalay
et al. (1968), Verloo (1974), Bunzl et al. (1976).

Frost and Griffin (1977) studied the influence of pH on retention of heavy metals. At pH > 5.5 retention of Cu increased considerably due to increasing adsorption of Cu-hydroxy complexes and formation of $Cu(OH)_2$. In clay minerals Kishk and Hassan (1973) also stated an increasing adsorption in function of pH. At low pH values adsorbed amounts of Cu were low, probably due to increasing competition of H ions and the release of octaedric bound Mg, Fe and Al from 2 : 1 clay minerals.

According to Farrah and Pickering (1976) clay surfaces show an apparent preference for Cu-hydroxy ions, resulting in a strong pH dependent selectivity. As well $CuOH^+$ (Menzel and Jackson, 1950) as $Cu_2(OH)_2^{2+}$ and $Cu(OH)_2^0$ (Baes and Mesmer, 1976) may be preferentially adsorbed species. According to Harmsen (1977) specific ion-exchange mainly occurs at low coverage of the adsorption sites. He showed that specific exchange adsorption of heavy metals was mainly due to adsorption of metal ions or first hydrolysis products on surface hydroxyls associated with organic matter or mineral surfaces.

The present paper discusses results of additional studies on the influence of pH, soil texture, organic matter content and the presence of other competing ions in the soil solution on the retention and release of Cu by soils. The distribution of added Cu over soluble, adsorbed and complexed forms was also studied in function of pH.

MATERIALS AND METHODS

Three different textured soils were sampled from the surface horizon (0 - 20 cm) and analytically characterised according to the procedures described by Cottenie et al. (1979). The main chemical characteristics of the selected soils are reported in Table 1.

TABLE 1

SOME CHEMICAL CHARACTERISTICS OF THE SELECTED SOILS

Location	Texture	CEC meq/100g	pH-H_2O	pH-KCl	% $CaCO_3$	% C	% clay
Peruwelz	light sandy loam	9.0	6.25	5.50	0.2	1.30	7.8
Ath	light loam	11.5	6.50	5.75	0	1.42	13.6
Pervijze	heavy clay	32.5	7.10	6.65	16.5	1.79	39.6

In a first experiment Cu adsorption isotherms were obtained by adding increasing amounts of Cu to a soil-water suspension (soil/solution ratio = 1/5). The time necessary for reaching equilibrium was previously studied and it was shown that 24 h of shaking were largely sufficient.

After shaking mechanically for 24 h, soil suspensions were centrifuged and the supernatants analysed for soluble Cu by atomic absorption spectroscopy. The quantities of adsorbed Cu were calculated by difference between the soluble amounts before and after equilibration. In a second experiment soils were suspended in a solution of 0.01 n $CaCl_2$ in order to study the effect of competing Ca ions on Cu adsorption.

In a third phase influence of pH on the distribution of added Cu in the soil was experimentally studied as follows: 50, 100 and 1 000 ppm (mg/kg soil) Cu were added to soil suspensions, which by means of appropriate HNO_3 additions were adjusted to different pH values, ranging from the original soil pH to pH 0.5. Each pH value was kept constant for 30 minutes by means of a pH-stat apparatus. Then suspensions were centrifuged and soluble Cu determined. After washing and drying the soil remaining in the centrifuge tube, the following Cu fractions were determined:
- exchangeable Cu by percolation with 1 n NH_4OAc pH = 7
- organically bound Cu by extraction with 0.2 n NaOH (Verloo, 1974).

RESULTS AND DISCUSSION

Adsorption experiments

Cu-adsorption isotherms for the three selected soils are
represented in Figure 1. These curves show the steepest slope
for the Peruwelz soil, indicating lower adsorptive capacities
of this soil. For the heavy textured soil of Pervijze a
flattening of the curve is observed, indicating a greater
buffering effect of the solid phase on Cu present in the soil
solution. Experimental data were fitted to the Langmuir
equation, which is given by:

$$\frac{C}{x/m} = C/b + 1/kb$$

where C is the equilibrium concentration of Cu(mg/l)

x/m the amount adsorbed (mg/kg soil)

b the adsorption maximum (mg/kg)

k the Langmuir bonding energy coefficient (ppm^{-1})

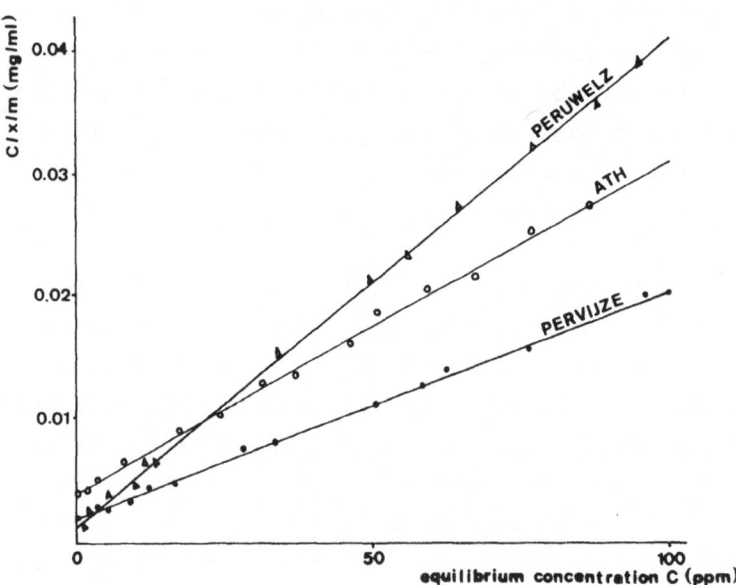

Fig. 1. Cu adsorption isotherms for the three selected soils.

The statistical significance of the correlation coefficient for C/x/m versus C was the criterion by which the fit of the data to the Langmuir isotherm was tested. A plot of C/x/m versus C yields a straight line with intercept 1/kb and slope 1/b, and permits to calculate the Langmuir coefficients b and k. The calculated values are given in Table 2.

TABLE 2

LANGMUIR PARAMETERS B AND K FOR THE EQUATION DESCRIBING Cu ADSORPTION BY THE SELECTED SOILS

Soil sample	adsorption maximum b mg/kg	% CEC	Bonding energy coefficient k (ppm^{-1})	Correlation coefficient r
Peruwelz	2 480	86.7	0.330	0.999***
Ath	3 845	105.2	0.334	0.997***
Pervijze	5 000	48.5	2.000	0.998***

The high values of the correlation coefficient r indicate that Cu adsorption conformed to the Langmuir equation. The influence of soil characteristics, such as pH, % clay, organic matter content and texture on Cu adsorption is illustrated by the fact that the highest values of b and k were obtained for the heavy soil of Pervijze.

The eventual effect of competition of other cations (Ca^{2+}) in the soil solution on Cu adsorption was studied by determining Cu adsorption isotherms in the presence of 0.01 n CaCl$_2$. The results obtained for the Peruwelz soil are shown in Figure 2, representing the percentage of added Cu adsorbed by the soil as a function of the amounts of Cu added, as well in 0.01 n CaCl$_2$ as in distilled water.

The experimental data were also fitted to the Langmuir adsorption equation and the calculated values are given in Table 3.

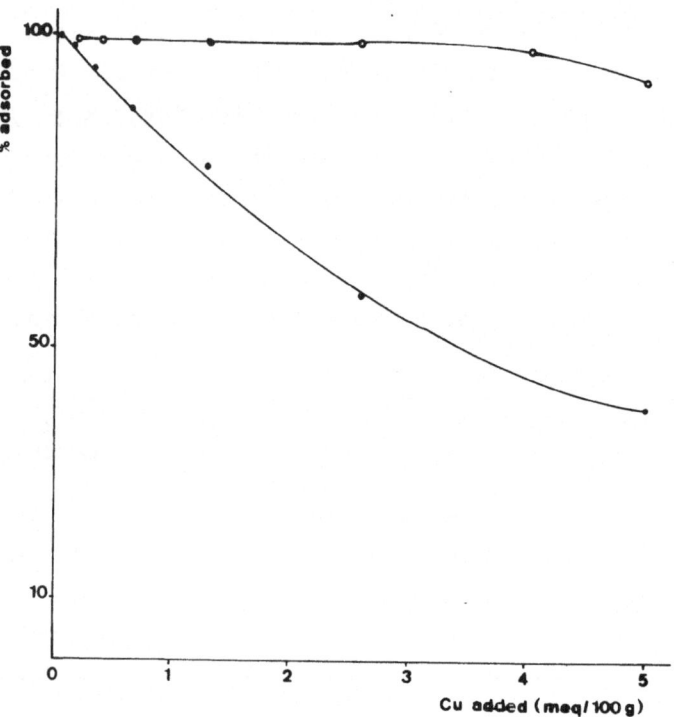

Fig. 2.   Percentages of added Cu adsorbed by the Peruwelz soil suspended
in water (o) and in 0.01 n CaCl$_2$(o)

TABLE 3

LANGMUIR PARAMETERS b AND k   FOR THE ADSORPTION OF Cu BY THE SOIL OF
PERUWELZ IN THE PRESENCE OF 0.01 n CaCl$_2$

Adsorptive capacity b		Bonding energy coefficient k (ppm^{-1})	Correlation coefficient r
ppm	% of CEC		
666	23.3	0.101	0.990 ***

The Langmuir parameters b and k were significantly lower
in the presence of 0.01 n CaCl$_2$, indicating that Ca^{2+} competes
effectively with Cu for the adsorption sites.   In this context,
Kiekens (1980) showed a different behaviour of the heavy metal
pairs Zn-Cd and Pb-Cu.   The influence of Ca^{2+} competition was
more pronounced for the elements Zn and Cd; which means that
mainly ion-exchange reactions take place.   The elements Pb and

Cu however may be more specifically bound, whereby organic matter plays an important role. It is sufficiently known that Cu and Pb form the most stable organo-mineral complexes with soil humic substances. Furthermore, as is evident from Figure 2, the percentage of added Cu adsorbed by the soil decreased with increasing amounts of Cu added. This agrees with the results obtained by Harmsen (1977), who showed that the selectivity coefficient of the Cu-Ca exchange reaction reaches a maximum at very low coverage of the adsorption complex with Cu ions, and decreases progressively at higher amounts of Cu added.

This may be explained by assuming the adsorption complex to consist of two types of adsorption sites, namely low and high selectivity sites. When adding Cu ions to the soil, first the high selectivity sites are occupied. When all these sites are exhausted, copper is adsorbed less strongly at low selectivity sites. Petruzelli and Guidi (1976) showed the existence of different kinds of linkages between copper and soil organic matter. Their results from gel chromatography suggest that higher molecular weight humic materials strongly link copper which becomes poorly available to plants. Plessis and Burger (1971) also found that Cu is less strongly adsorbed at high site coverage.

Distribution of added Cu in the soil as a function of pH

The observed distribution of added Cu over soluble (S), exchangeable (A), complexed (C) and 'fixed' (F) forms as a function of pH of the soil suspension is shown in Figure 3. The fractions A and C are related to the amounts of Cu retained by the soil, as they were determined on the air-dried soil after determination of the soluble quantities (S). The fraction F was calculated as the difference between the retained amounts and the fractions A and C.

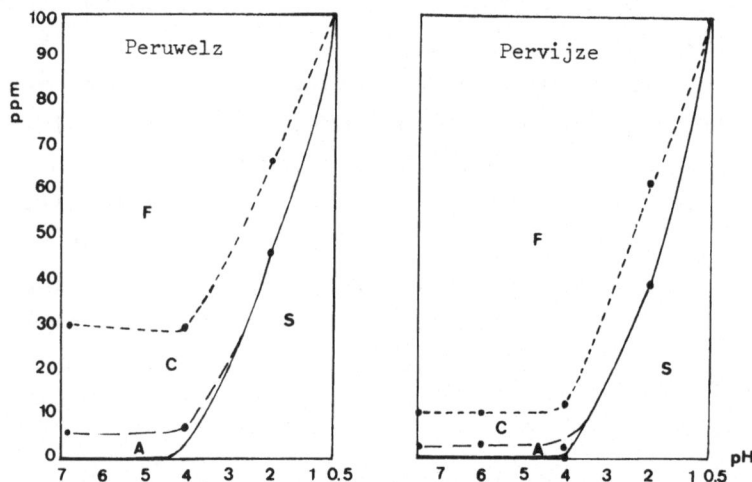

Fig. 3. Distribution of added Cu (100 ppm) in function of pH over soluble (S) exchangeable (A), complexed (C) and 'fixed' (F) forms.

At the actual soil pH, added Cu is almost completely retained by the soil. Decreasing the pH of the soil suspension resulted in a slow increase of the soluble quantities until pH values of about 3 - 4. At pH < 3 - 4 desorption and solubilisation of added Cu became quantitatively important and was practically complete at pH = 0.5.

The fraction of added Cu retained by the soil was not quantitatively exchangeable with 1 n NH_4OAc pH = 7, indicating that Cu retention was partly irreversible. The irreversible character of Cu retention is also influenced by soil characteristics: indeed, the fraction of retained Cu exchangeable with 1 n NH_4OAc pH = 7 was quantitatively less important for the heavy clay soil from Pervijze.

The fraction of retained Cu extractable with 0.2 n NaOH was more important in the light textured soil than in the heavy clay soil. This means that in the heavy clay soil ion-exchange and precipitation mechanisms are mainly responsible for Cu retention; while in the Peruwelz soil, with a lower clay content and pH, interactions with organic substances are relatively more important.

Influence of organic matter on the behaviour of Cu

As already mentioned, soil organic matter may form organo-mineral complexes with heavy metals, and particularly with Cu. As a function of their solubility, soil humic substances can be separated in humic acids, only soluble in alkali and fulvic acids which are soluble in both alkali and acid. Fulvic acids form soluble metal chelates over a wide pH range, increasing the solubility of Cu. An important fraction of Cu present in the soil solution and in surface waters appears to be bound to a yellowish compound with fulvic acid properties.

Hodgson et al. (1965) estimated that 76 - 99.5% of Cu present in the soil solution was linked to soil organic constitutents. Delas (1967) observed that Cu ions form chelates preferentially with the most mobile fractions of organic matter, i.e. with fulvic acids. Verloo (1979) observed that 93% of soluble Cu was present as fulvate complexes in a soil leachate. In contrast, complexes of heavy metals with humic acids are less soluble, especially in acid medium.

Though the composition of soil organic matter is rather heterogeneous, several authors have tried to determine the stability constants of organo-copper complexes:

1) According to Coleman et al. (1956) the stability constant log K of the Cu-peat complex equals 6.5.

2) Schnitzer and Skinner (1966) found the following values for Cu fulvates:
$$\log K = 5.78 \text{ at pH } 3.5$$
$$= 8.69 \text{ at pH } 5.0$$

3) Courpron (1967) found a log K value = 3.23 at pH 3.5 for Cu-fulvate, and log K = 7 at pH = 5 for Cu-humate.

According to Verloo (1974, 1979) stability constants are only meaningful for really soluble complexes, and do not give information about the behaviour of insoluble humate complexes. Therefore, he suggests the use of a stability index, defined as:

$$SI = \frac{\Sigma \ (M \ organic)}{\Sigma \ (M \ inorganic)}$$

where Σ(M organic) is the sum of the concentrations of a cation M complexed with organic matter, and Σ(M inorganic) is the sum of the concentration of all inorganic forms of M in equilibrium with organic forms.

Verloo (1974) determined experimentally the stability indexes of Cu-humic acid and Cu-fulvic acid complexes as a function of pH. The stability index for Cu-HA increased with increasing pH and reached a maximum value of 9.15 at pH = 4. This means that 9.15 times more Cu was in organic form (HA) than in an inorganic one. Furthermore most of complexed Cu was in insoluble form at this pH value; which means that Cu is fixed by organic matter. On the contrary at pH values > 6 Cu-humates were soluble with a maximum stability at pH 9. The stability indexes of Cu-fulvic acid complexes (Cu-FA) were much lower than those of the corresponding Cu-HA complexes over the whole pH range. Thus it may be concluded that Cu will mainly be bound by humic acids, and, depending on the pH, the Cu-HA complex will be soluble (at high pH) or insoluble (at low pH), when increasing the electrolyte concentration of the system by addition of Ca e.g., soluble humates flocculate and only fulvates remain in solution.

Verloo and Cottenie (1972) and Verloo (1974) determined the free and complexed forms after equilibration at different pH values and Figure 4 shows the percentage distribution of Cu in purified humic acid copper systems.

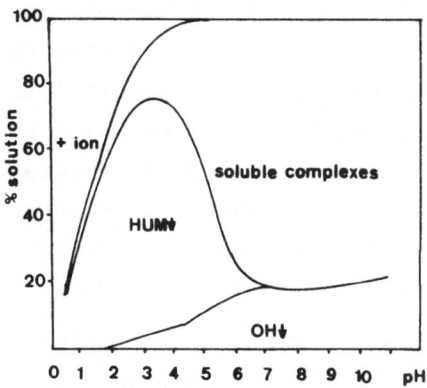

Fig. 4. Percentage distribution of Cu in purified humic acid copper systems at different pH-values.

At pH > 6.5 copper is present as a soluble humate complex and as a $Cu(OH)_2$ precipitate. At pH values lower than 6.5 a precipitation of an insoluble Cu-humate is observed besides a decreasing fraction of $Cu(OH)_2$. The stability of the soluble humate complex strongly decreases and disappears completely at pH = 1. In strong acid solution Cu is mainly present as free ion Cu^{++} and as an insoluble humate complex. At pH 3.5, representing the typical value of an acid peat, some 75% of total Cu is immobilised as an insoluble humate complex. The soluble Cu humate complex shows a maximal stability at pH 8, at which 83% of total Cu is present in solution as a humate complex.

When the previous results are transferred to a soil system, Cu-organic matter interactions can be described as follows (Verloo, 1979):

- in acid soil systems, pH 4 to 5, Cu is mainly fixed in the insoluble humic acid fraction. When this fraction is quantitatively important and the mineralisation rate is low, e.g. in peat soils, large amounts of Cu may be immobilised and may cause Cu deficiency to plants

- from pH 6 to 7, stability of Cu-humate complexes decreases, while Cu-fulvate complexes show a maximal stability. Depending on the electrolyte concentration of the soil solution, Cu-humates will be soluble or flocculated. Soluble humates may migrate in the profile to zones with higher electrolyte concentrations or lower pH, where they will accumulate. Cu may be released upon mineralisation of organic constituents

- at higher pH, behaviour of Cu-organic matter compounds in the soil will be completely governed by the presence of Ca as a flocculating agent of the soluble Cu-humates.

CONCLUSIONS

It may be concluded that soil organic matter and mineral soil colloids are important factors affecting the behaviour of Cu in soils and sediments. At normal soil pH values, the adsorbed fraction of Cu is much higher than the water soluble one. Desorption occurs at lower pH and its critical value is about 3 for copper. An important fraction of Cu may be involved in specific adsorption reactions, which is manifested in a selective uptake. Furthermore a quantitatively important fraction may be irreversibly fixed, the extent of which seemed to be greater at low occupancy of the complex with Cu ions. The fulvic acid fraction of soil organic matter will mainly form soluble complexes with Cu and hence increases its mobility. On the contrary humic acids mainly form insoluble complexes and can be considered as the organic storage place for Cu.

REFERENCES

Baes, C.F. and Mesmer, R.E., 1976. The hydrolysis of cations. Wiley-
Interscience, New-York.

Bingham, F.T., Page, A.L. and Sims, J.R., 1964. Retention of Zn and Cu
by H-montmorillonite. Soil Sci. Soc. Am. Proc., 28, 351-354.

Bower, C.A. and Truog, E., 1941. Base exchange capacity determination as
influenced by nature of cation employed and formation of basic
exchange salts. Soil Sci. Am. Proc., 5, 86-89.

Bunzl, K., Schmidt, N. and Sansoni, B., 1976. Kinetics of ion exchange in
soil organic matter. IV. Adsorption and desorption of Pb^{2+}, Cu^{2+},
Zn^{2+} and Ca^{2+} by peat. J. Soil Sci., 27, 32-41.

Cavallaro, N. and Bride Mc., M.B., 1978. Copper and cadmium adsorption
characteristics of selected acid and calcareous soils. Soil Sci.
Soc. Am. J., 42, 550-556.

Coleman, N.T., McLung, A.C. and Moore, D.P., 1956. Formation constants for
copper-peat complexes. Science, 123, 330.

Cottenie, A., Verloo, M., Velghe, G. and Kiekens, L., 1979. Analytical
methods for plants and soils. Lab. Analyt. & Agrochemie, R.U.G.
pp. 39.

Courpron, C., 1967. Détermination des constantes de stabilité des complexes
organo-métalliques des sols. Ann. Agron., 18, 623.

Davies, R.I., Cheshire, M.V. and Graham-Bryce, I.J., 1969. Retention of
low levels of copper by humic acid. J. Soil Sci., 20, 65-71.

Delas, J., 1967. Liaisons entre cuivre et matière organique dans un podzol
développé sur sable des Landes et accidentellement enrichi en cuivre.
Ann. Agron., 18, 17.

DeMumbrum, L.E. and Jackson, M.L., 1956. Infrared adsorption evidence on
exchange reaction mechanisms of Cu and Zn with layer silicates and
peats. Soil Sci. Soc. Am. Proc., 20, 334-337.

Elgabaly, M.M. and Jenny, H., 1943. Cation and anion interchange with zinc
montmorillonite clays. J. Phys. Chem., 47, 399-408.

Farrah, H. and Pickering, W.F., 1976. The sorption of Cu species by clays.
II. Illite and montmorillonite. Aust. J. Chem., 29, 117-184.

Frost, R.R. and Griffin, R.A., 1977. Effect of pH on adsorption of copper,
zinc and cadmium from landfill leachate by clay minerals. J. Environ.
Sci. Health, A12, 139-156.

Guidi, G., Petruzelli, G. and Sequi, P., 1972. Interazioni tra sostanza
organica e microelementi. Atti IX Simposio Internazionale di
Agrochimica, Punta Ala, 407-411.

Harmsen, K., 1977. Behaviour of heavy metals in soils. Agric. Res. Rep.,
866, Centre for Agric. Publ. and Doc., Wageningen.

Hodgson, J.F., Geering, H.Z. and Norvell, W.A., 1965. Micronutrient cation
complexes in soil solution : Partition between complexed and
uncomplexed forms by solvent extractions. Soil Sci. Soc. Am. Proc.,
29, 665.

Kiekens, L., 1980. Adsorptieverschijnselen van zware metalen in gronden.
Doct. Dissertation, R.U.G.

Kishk, F.M. and Hassan, M.N., 1973. Sorption and desorption of copper by
and from clay minerals. Plant and Soil, 39, 497-505.

Levi-Minzi, R., Soldatini, G.F. and Riffaldi, R., 1976. Cadmium adsorption
by soils. J. Soil Sci., 27, 10-15.

McLaren, R.G. and Crawford, D.V., 1973. Studies on soil copper. II. The
specific adsorption of copper by soils. J. Soil Sci., 24, 443-452.

Menzel, F.G. and Jackson, M.L., 1950. Mechanism of sorption of hydroxy-
cupric ion by clays. Soil Sci. Soc. Am. Proc., 15, 122-124.

Misra, S.G. and Tiwari, R.C., 1966. Retention and release of copper and
zinc by some Indian soils. Soil Sci., 101, 465-471.

Petruzelli, G. and Guidi, G., 1976. Influence of soil organic matter on
copper availability to plants. Z. Pflanzenernaehr. Bodenkd., 6,
679-684.

Petruzelli, G., Guidi, G. and Lubrano, L., 1978. Organic matter as an
influencing factor on copper and cadmium adsorption by soils. Water,
Air and Soil Pollution, 9, 263-269.

Plessis, S.F. Du and Burger, R. Du T., 1971. Die spesifiekte adsorpsie van
Koper deur Kleiminerale en grondfraksies. Agrochemophysica, 3, 1-10.

Schlichting, E., 1955. Kupferbindung und -fixierung durch Huminsäuren. Z.
Pfl. Ernähr. Dung. Bodenkd., 69, 134-137.

Schnitzer, M. and Skinner, S.I.M., 1966. Stability constants of Cu^{++},
Fe^{++} and Zn^{++}-fulvic acid complexes. Soil Sci., 102, 361.

Shuman, L.M., 1975. The effect of soil properties on zinc adsorption by
soils. Soil Sci. Soc. Am. Proc., 39, 454-458.

Szalay, A. and Szilagyi, M., 1968. Laboratory experiments on the retention
of micronutrients by peat humic acids. Plant and Soil, 29, 219-224.

Udo, E.J., Bohn, H.L. and Tucker, T.C., 1970. Zinc adsorption by
        calcareous soils. Soil Sci. Soc. Am. Proc., 34, 405-407.

Verloo, M.G. and Cottenie, A., 1972. Stability and behaviour of complexes
        of Cu, Zn, Mn and Pb with humic substances of soils. Pedologie,
        XXII, 174-184.

Verloo, M.G., 1974. Komplexvorming van sporenelementen met organische
        bodemkomponenten. Doct. Dissert., R.U.G.

Verloo, M.G., 1979. Influence of soil organic matter on the behaviour of
        heavy metals in soils and sediments. In : Essential and non
        essential trace elements in the system soil-water-plant., Lab.
        Analyt. & Agroch. R.U.G.

DISCUSSION

G.A. Fleming *(Ireland)*

In the first two soils that you spoke about, the heavy
soil and the light soil, what was the organic matter content
of one versus the other?

L. Kiekens *(Belgium)*

There was not much difference; the heavy soil was 1.7 and
the light soil 1.5, I think from memory.

P.H.T. Beckett *(UK)*

I agree with your point of there being two kinds of site
for retaining copper but if this is so, why is Figure 1 linear?
One would expect it to be curved at the lower end.

L. Kiekens

I think the reason is that we did not have enough
observations to divide it into two linear parts. At the lower
concentrations it is necessary to make many more observations
in order to obtain two parts.

THE APPLICATION OF COPPER IN SEWAGE SLUDGE AND PIG MANURE TO AGRICULTURAL LAND IN ENGLAND AND WALES

R.J. Unwin

Ministry of Agriculture, Fisheries and Food,
Great Westminster House, Horseferry Road, London, UK

ABSTRACT

The natural copper content of soils in England and Wales extractable in EDTA is usually < 5 mg/l and consequently most soils can tolerate additions up to 280 kg/ha Cu without significantly limiting the cropping potential of the land. This guideline may be used for the application of sewage sludge or pig manure. Although most fattening pigs receive copper supplements the quantity of copper recycled will vary according to management practices. Only a little over 1/3 of the pig herd is stocked at rates which would result in average copper additions to soil exceeding 2 kg/ha per year. The area of land receiving more than 5 kg/ha Cu per year is small. Advice can be given which will prevent harmful accumulations in soil and minimise the risks to grazing livestock. Reducing the permitted level of copper supplementation will not remove the environmental hazards but sensible disposal practice can achieve this. Authorities are increasingly tightening controls on pollution of water and air from pig units. The recent trend for a greater proportion of the national herd to be kept on holdings with adequate land for manure disposal is expected to continue.

COPPER IN SOILS AND AGRICULTURAL PRODUCTION

Although copper occurs most commonly in the natural state as oxides or sulphides much of the agronomically important fraction in soils is complexed with organic matter to form stable associations. The proportion of the metal extracted from soil using 0.05 M NH_4 - EDTA (EDTA Cu) has been found to be satisfactory in the United Kingdom for assessing the ability of soil to supply copper for crop growth.

Archer (1980) reported copper contents of soils collected from randomly selected farms in England and Wales including land which might have received additions of copper. Total copper (nitric/perchloric digest) ranged 2 - 195 mg/kg air dried soil, median 17 mg/kg (for 751 samples) and EDTA Cu 0.5 - 74 mg/l air dried soil, median 4.4 mg/l (662 samples). For all samples 95% had < 40 mg/kg total Cu and 92% < 10 mg/l EDTA Cu. When grouped into 14 soil parent material classes the median EDTA Cu for each class varied from 3.1 - 6.5 mg/l except for fen peats 7.5 mg/l, alluvial soils 7.0 mg/l and sandstone soils 2.3 mg/l. The data indicate on average 25% of the copper to be EDTA extractable. This is rather higher than the 10 - 20% reported for Scottish soils (Williams, 1980).

Copper deficiency in crops in England and Wales has been discussed by Caldwell (1976) and reported to occur mainly on three groups of soils: peats, podsolic sands and shallow humose chalk soils. An EDTA Cu of < 2.5 mg/l may cause deficiency depending on other factors, notably pH and organic matter. Grass very rarely exhibits copper deficiency but with a mean content in leafy grass of 9.0 mg/kg Cu in dry matter (range 2 - 15 mg/kg Cu, s.d ± 3.67, ADAS, 1975), the dietary requirements of grazing stock may sometimes not be met. Mineral supplementation or copper therapy is often necessary.

Therefore in England and Wales most soils are adequately supplied with copper for crop growth. However these soil levels are often not high enough to meet the copper requirements of

grazing livestock particularly under modern management systems
of pure grass swards and high fertiliser use. Environmental
problems are possible as a result of applications to land of
sewage sludge or pig manure contaminated with high concentrations
of copper. However they can be avoided by the sensible management of such applications.

COPPER CONTENT OF SEWAGE SLUDGE AND APPLICATION TO LAND

Agricultural land receives 40% of the sewage sludge (on a
dry solids basis) produced in England and Wales. The copper
content of a range of sludges has been reported as 36 - 2 889 mg/kg
Cu on dry solids, the median of the 2 379 samples being
546 mg/kg Cu (DOE, in press).

Berrow and Burridge (1980) reported analytical data when
heavy dressings of sludge had been applied to horticultural land.
The proportion of added copper detected in the treated soil as
EDTA Cu was variable at one site (mean 50%) and fairly constant
around 34% at the other. There was no evidence of a significant
change of extractability during the eight years following
application (Berrow, in press).

Richardson (1980 and unpublished data) has reported total
and EDTA Cu analyses on 86 soils which had received varying types
and rates of sewage sludge over different lengths of time. Total
soil Cu varied 9 - 570 mg/kg. In only 10% of samples was the EDTA
Cu fraction below 30% of the total value. For soils with < 6.8%
organic matter, on average 60% of the copper was extractable in
EDTA but at low levels of total copper (< 70 mg/kg Cu) there
was a wide variation in extractability (10 - 95%). The soils with
the highest total copper levels also had high organic matter
contents (8 - 34%) and the extractability of copper from these
was only 30 - 40%.

Evidence from field trials on the phytotoxicity of copper
in soils is limited and results from pot trials, especially
those using metal salts, must be treated with caution.

Williams (1975) has reviewed the available information and concluded that whilst the most sensitive crops may be affected at
50 mg/kg EDTA Cu at a slightly acid pH, more resistant species
would not be affected until 100 mg/kg EDTA Cu had been reached.
He stressed the importance of pH in determining crop susceptibility.

Results from two field trials to determine the phytotoxic
effects of zinc, copper and nickel applied in sewage sludge were
reported by Marks et al., (1980). Statistical transformation of
their results suggested that in a sandy loam (pH 6.0 - 6.2,
organic matter 1.8%) a 10% reduction in yield of red beet would
have resulted at 132 mg/kg EDTA Cu (standard error ±98). In a
silt loam (pH 6.2 - 7.0) the beet could tolerate twice as much
copper. Lettuce was found to be less sensitive than red beet
whilst celery was not affected by an application of 755 kg/ha Cu.

In 1971 the Agricultural Development and Advisory Service
(ADAS, 1971) recommended that the maximum permissible level of
copper addition to previously uncontaminated soil in sewage
sludge should be 280 kg/ha Cu. This would be reduced if zinc
or nickel were present in the sludge. When incorporated into
the top 20 cm of soil this addition will raise total copper by
140 mg/kg and the EDTA Cu by approximately 70 mg/kg which, it
is considered, is a realistic safeguard for crops at pH 6.5.
This recommendation is included in guidelines for the disposal
of sewage sludge to land (DOE, 1977)

Where the pH is likely to be below 6.5, ADAS would recommend
a lower limit. For grassland where the copper and other metals
accumulate in the surface soil lower additions may be desirable
especially if the sward is not ploughed at intervals. This is
especially necessary as the pH of grassland is often below 6.0
and the metal burden might then affect clovers even if the more
tolerant grasses were unaffected. Higher rates would be
considered safe on naturally calcareous soils.

Studies on the transormation of copper in soil following
sludge application, the suitability of different extractions and

experiments to establish phytotoxic levels in soils are in progress. The guidelines outlined above may need to be modified according to the results.

COPPER CONTENT OF PIG SLURRY AND LAND APPLICATION

The copper content in slurry from pigs receiving a copper supplement diet is often in the range 600 - 900 mg/kg Cu in dry matter. The level can fluctuate considerably in samples taken at different times from the same storage facility e.g. 273 - 1 990 mg/kg Cu for 34 samples taken over a four year period (Unwin, 1980).

Sluijsmans (1978) recommended that following pig manure application HNO_3 Cu should not exceed 50 ppm in arable top soil. This was stated to be equivalent to 40 ppm EDTA Cu but the methods were not given. On the evidence from sewage sludge trials as discussed above this arable soil limit cannot be justified for soils maintained at pH 6.5. In order to limit the ingestion of copper by stock he suggested that copper in the 0 - 5 cm soil layer should not exceed 15 - 20 ppm HNO_3 Cu where sheep might graze but that 100 ppm was acceptable for cattle. A total addition of 150 kg/ha Cu was suggested to limit the HNO_3 Cu level to < 50 ppm.

Research into the availability to stock of copper in different feedstuffs and from soil and slurry is continuing in the UK. Available information does not support the very restrictive soil copper limit proposed for sheep pastures and this subject must be considered as a matter of importance. No proven cases of copper toxicity to sheep at grass on commercial holdings exposed to sludge or slurry have been encountered by ADAS in England and Wales which is consistent with experimental results such as those of Gracey et al. (1976). They reported no danger to health of sheep continually grazing permanent pasture for 3 years during which time 900 m^3/ha of copper rich slurry was applied. In contrast Kneale and Smith (1977) reported ewes

close to death when forced to graze contaminated pasture but in
this case 500 m^3/ha pig slurry was applied in one year. When
the sheep only grazed during spring or summer following winter
application of 250 m^3/ha pig slurry blood copper levels were
only slightly raised.

Studies on the biological availability of copper rich soil
are in progress (Suttle, private communication) but the ingestion
of untreated soil has been shown to lower the absorption by
sheep of herbage copper (Suttle et al., 1975). Copper in pig
slurry has been shown to be of similar absorbability to other
dietary sources, (Dalgarno and Mills, 1975; Suttle and Price, 1976)
but is likely to be influenced by other constituents, particularly
molybdenum and sulphur (Suttle and Price, 1976; Suttle, 1977).

Soil ingestion by sheep during spring and summer is < 5%
of dry matter intake so that even at 50 ppm HNO_3 Cu in soil only an
extra 2.5 mg per day of this copper fraction will be ingested
compared to an expected 7 - 10 mg Cu from herbage alone. For the
reasons outlined above it is possible that the net effect of such
soil ingestion would be a reduction in the amount of copper absorbed.

The use of soil analysis to monitor copper accumulations in
soil must be undertaken with care over an extended period of time.
Since 1969 ADAS have taken samples regularly from a commercial
pig fattening and dairy farm (Batey et al., 1972; Unwin, 1980).
Although soil and herbage copper contents have tended to incease,
individual values from the 11 enclosures have fluctuated widely.
Among the reasons for the fluctuation has been uneven application
within enclosures leading to increased sampling error, ploughing
to reseed swards and a tendency for EDTA Cu to decrease with time
between slurry applications. These factors must be considered
when monitoring soils for accumulations of copper.

Under field trial conditions on a silty clay soil it was
possible to account for virtually all of the copper applied in
pig slurry within the soil profile (Unwin, 1980). A maximum of
212 kg/ha Cu was applied and 30 - 45% was detected as EDTA Cu

within the cultivated layer. On this soil extensive cracking
in summer was considered responsible for an observed trans-
location of copper into the subsoil. Further evidence of the
more usual accumulation of copper in surface layers is available
(Table 1) from a lysimeter experiment with a sandy loam soil
previously described by Hewgill and Le Grice (1976).

TABLE 1

ACCUMULATION OF COPPER AFTER 3 ANNUAL APPLICATIONS OF PIG SLURRY TO A
SANDY SOIL IN LYSIMETERS (mg/kg EDTA Cu)

Soil layer	Control	1 800 m^3/ha pig slurry
Surface Mat	3.3	109
0-5 cm	1.9	23.3
5-15	1.4	3.8
15-30	1.0	1.5
30-45	0.5	0.7
45-60	0.5	0.5

Concern is often expressed about the effect of copper on
earthworm populations. Van Rhee (1975) reported that average
populations were least on 'poor class organic sandy soils',
where copper levels were highest after 10 years application of
pig waste. In contrast, Unwin (1980) reported a doubling of
numbers on plots receiving pig slurry containing the equivalent
of 212 kg/ha Cu compared to controls receiving only inorganic
fertilisers during the 4 year experiment.

COPPER CONTENT OF HERBAGE AND CROPS FOLLOWING PIG SLURRY
APPLICATIONS

The significance of copper contaminated pasture for
grazing stock is the subject of other papers at this meeting.
The following comments relate specifically to work undertaken
by ADAS. Some has previously been discussed elsewhere
(Unwin, 1980).

In field trials when heavy dressings of pig slurry were applied monthly throughout the year, but only to grass stubble during the growing season, copper levels in herbage were variable but first cuts in spring were always high (often 40 mg/kg Cu in dry matter). Physical contamination with slurry was possible but it was also shown in a glass house experiment that copper from slurry can be retained by grass after the obvious physical contamination is removed by washing prior to analysis.

At the second cut copper contents were much lower but tended to rise in some years as the season progressed. This has also been observed in continuing studies with soil removed from the experiment and used for a replicated experiment in small plots. One year after the last slurry application an autumn cut contained 40 mg/kg Cu in dry matter but as was observed in the earlier trial the effect decreased in subsequent years.

TABLE 2

COPPER CONTENT OF GRASS (mg/kg IN DRY MATTER) GROWN ON SOIL PREVIOUSLY SUBJECTED TO HEAVY DRESSINGS OF PIG SLURRY. 6 CUTS PER SEASON.

	Soil copper mg/l EDTA Cu	1977 Mean	1977 Max	1978 Mean	1978 Max	1979 Mean	1979 Max
Control	2	9.7	13.9	10.6	16.9	11.4	14.3
High slurry	42	18.2	40.2	15.6	31.1	12.7	13.9

When the rate of slurry in the field trial approximately balanced the major nutrient requirements of the sward an average 2.5 kg/ha Cu per year was applied. At no time did grass on this monthly spreading regime contain more than 15 mg/kg Cu in dry matter. The mean copper content for the season was only slightly above that of the control (Table 3). At the highest rate copper levels were no higher in the fourth year of application than the first.

TABLE 3

COPPER CONTENT OF HERBAGE GROWN ON PLOTS RECEIVING MONTHLY DRESSINGS OF
PIG SLURRY FOR 4 YEARS. (MEAN OF 6 CUTS).

Mean copper addition in slurry kg/ha Cu per year	Mean herbage copper content mg/kg Cu in dry matter	
	1973	1976
Control	7.5	6.5
2.5	7.9	6.9
53	13.3	13.3

The behaviour and effects of copper added to soil in pig
manure is similar to that applied in sewage sludge with the
possible exception of a decrease of EDTA extractability with
time. It is therefore considered appropriate to adopt the same
guidelines for levels of copper addition as have been described
for sewage sludge.

THE PIG INDUSTRY IN ENGLAND AND WALES

The pig herd in England and Wales has fluctuated around
6.5 million head in recent years, 11 - 12% of which are breeding
stock. They are kept on some 30 000 holdings which comprise
approximately 1.8 million hectares of crops and grass or 19% of
the total crop and grass area. Data on stock numbers and land
area are collected in the annual MAFF census of all holdings, the
most recently published full analysis being for 1975 (MAFF, 1978).
Surveys giving an indication of the structure and management of
the industry have been made by the Meat and Livestock Commission
(MLC, 1979) and the National Farmers Union (NFU, 1980). These
sources suggest that 4 out of 5 producers breed and fatten pigs
on the same premises. Less than 5% of herds are fed on waste
materials, the rest on compound feeds. Official statistics on
the extent of copper supplementation are not available, but a
1974 survey in Scotland indicated 78% of all fattening herds
received added copper. The proportion in England and Wales is
likely to be at least as high and might be expected to have

increased as the benefits in pig performance have been more widely accepted. The proportion of total pigs treated is likely to be higher than the proportion of herds.

In an Agricultural Research Council co-ordinated trial at 19 centres the mean copper intake of pigs fed with 200 mg/kg added copper was 18.96 g Cu per pig up to 53 kg liveweight and 42.75 g Cu up to slaughter at 90 kg liveweight. (Braude, private communication). This is similar to the '38 g Cu per pig delivered', used by Sluijsmans (1978) in a discussion on copper recycling, although a lower level of supplementation was assumed. When assessing the likely copper loading to land not only the level of supplementation but the slaughter weight and growth rates must also be considered.

It is confidently expected that in the future a greater proportion of pigs will be kept in larger herds on arable farms. This change will be accelerated by the need to reduce water and odour pollution and planning controls which will require adequate land to be available for manure disposal.

AVOIDING ENVIRONMENTAL COPPER PROBLEMS FROM PIG MANURE

As a general guideline to avoid phytotoxicity problems the available evidence indicates that the addition limits proposed for sewage sludge are also appropriate for applications of pig manure, i.e. 280 kg/ha Cu subject to reduction when appreciable zinc is present in the manure. This amount should not be applied in less than 30 years and metal accumulations in the surface layer of grassland soils should be avoided by cultivations or a reduced limit.

In order to avoid general environmental problems farmers in England and Wales are advised to limit manure applications to rates which will not exceed the nutrient requirements of crops. A survey of the most intensively stocked pig units in England and Wales in 1977 (MAFF, unpublished data) indicated that the great majority of farmers follow this advice, disposing of surplus manure on neighbours' land where necessary. When this advice is followed applications of copper in pig manure are unlikely greatly to exceed 2.5 kg/ha Cu per year. In these circumstances it would be many years before the advised limit was reached, whilst the level of copper uptake into herbage will remain below that which is toxic to grazing sheep. Government farm census data (see Appendix I) and the above mentioned survey of intensive units indicate that the area of land currently receiving more pig slurry than crops need and which might require frequent monitoring for copper contamination may be no more than 2 000 - 3 000 ha.

It is undesirable for sheep to graze grass or forage crops physically contaminated with copper rich pig slurry, or to feed in fields in such a way that they may ingest significant quantities of soil contaminated with such manure before it has been incorporated by cultivation. However, very few sheep are kept on farms with intensive pig units and farmers should not have difficulty in following this advice. In practice no cases of copper toxicity in sheep from this cause have been reported in England and Wales even when these precautions have been ignored.

In the practical farming situation the safe disposal of copper from pig slurry has not been a problem in the UK. Advice can be given to ensure that this remains so. In view of the varying quantities of copper fed to pigs and differences in slaughter weight and feed conversion efficiency, reducing the level of copper supplementation from 200 to 125 ppm will not prevent environmental problems. This can only be achieved by sound practical advice relative to the individual situation and this approach is capable of controlling the extra copper resulting from the higher supplementation. On many farms some increases in herbage copper contents resulting from pig manure application may be of benefit to stock.

SUMMARY

1.   Information on the range of total and EDTA extractable
copper in soils is available for England and Wales. The majority
have < 5 mg/l EDTA Cu and over 90% < 10 mg/l EDTA Cu.

2.   Environmental problems are possible as a result of
applications to land of sewage sludge or pig manure contaminated
with high concentrations of copper. Reductions of the permitted
levels of copper supplementation in pig feeds will not eliminate
these problems but they can be avoided by sensible management
practices.

3.   The majority of fattening pigs in the UK are fed a copper
supplemented diet. However the area of land where pig stocking
rates and manure disposal practices might result in excessive
applications of copper is small and could easily be monitored.

4.   The best evidence available indicates that at pH 6.5
virtually all crop plants can tolerate 70 mg/l EDTA Cu in soil
and many substantially more. In calcareous soils the limit is
higher but at more acid pH crop sensitivity increases.

5.   The proportion of copper applied to soil in sewage sludge
or pig manure which may subsequently be extracted by EDTA can
vary according to soil, but is usually around 50 - 60%. The
extractability of sludge applied copper remains fairly constant
with time, but in the few cases studied by ADAS to date copper
applied in pig slurry has become rather less extractable during
the first few years after applications ceased.

6.   Copper content of herbage is raised by the application
of copper containing slurry. Annual dressings which do not
exceed the major nutrient requirements of grass are unlikely
to produce herbage copper levels which will affect grazing
stock provided they are applied after cutting or grazing. The
availability of copper for crop uptake falls when fresh slurry
dressings are withheld.

7.   Farmers are advised to monitor soils and crops for copper
when pig slurry is applied regularly to land. Advice can then
be given for future applications according to particular
circumstances.

REFERENCES

ADAS, 1971. Permissible levels of toxic metals in sewage used on
        agricultural land. MAFF ADAS Advisory Paper No 10.
ADAS, 1975. The important mineral elements in animal nutrition and their
        optimum concentration in forages. MAFF ADAS Advisory Paper No 16.
ADAS, 1976. Trace element deficiencies in crops. MAFF ADAS Advisory
        Paper No 17.
Archer, F.C., 1980. Trace elements in soils in England and Wales. In:
        Inorganic Pollution and Agriculture. MAFF Reference Book 326. HMSO.
Batey, T., Berryman, C. and Line, C., 1972. The disposal of copper-
        enriched pig manure slurry on grassland. J Br Grassld Soc. 27,
        139-143.
Berrow, M.L., in press. Trace elements in contaminated soils and related
        crops. Seminar on Progress in Crop Trials, December 1979. Department
        of the Environment, London.
Berrow, M.L. and Burridge, J.C., 1980. Trace element levels in soils:
        effects of sewage sludge. In: Inorganic Pollution and Agriculture.
        MAFF Reference Book 326. HMSO.
Dalgarno, A.C. and Mills, C.F., 1975. Retention by sheep of copper from
        aerobic digestion of faecal slurry. J. agric. Sci., Camb. 85, 11-18.
DOE, 1977. Report of the working party on the disposal of sewage sludge
        to land. Department of the Environment, London.
DOE, in press. Report of the sub-committee on the disposal of sewage
        sludge to land. Department of the Environment, London.
Gracey, H.I., Stewart, T.A., Woodside, J.D. and Thompson, R.H., 1976.
        The effect of disposing high rates of copper-rich pig slurry on grass-
        land on the health of grazing sheep. J. agric. Sci., Camb. 87, 617-623.
Hewgill, D. and Le Grice, S., 1976. Lysimeter study with pig slurry. In
        Agriculture and Water Quality. MAFF Technical Bulletin 32. HMSO.
Kneale, W.A. and Smith, P., 1977. The effects of applying pig slurry
        containing high levels of copper to sheep pastures. Experimental
        Husbandry 32, 1-

Marks, M.J., Williams, J.H. and Chumbley, C.G. 1980. Field experiments testing the effects of metal contaminated sewage sludges on some vegetable crops. In: Inorganic Pollution and Agriculture. MAFF reference book 326. HMSO.

MAFF, 1978. Agricultural Statistics England and Wales 1975. MAFF. HMSO.

MLC, 1979. Regional hit count, March 1979. Meat and Livestock Commission. London.

NFU, 1980. Who does what in pig production? British Farmer and Stockbreeder, 5 July 1980, pp 36-37.

Richardson, S.J., 1980. Composition of soils and crops following treatment with sewage sludge. In: Inorganic Pollution and Agriculture. MAFF Reference Book 326. HMSO.

Sluijsmans, C.M.J., 1978. The spreading of animal excrement on utilised agricultural areas of the Community. Commission of the European Communities Information on Agriculture No 47.

Suttle, N.F., 1977. Reducing the potential toxicity of concentrates to sheep by the use of molybdenum and sulphur supplements. Anim. Feed Sci. Technol. 2, 225-266.

Suttle, N.F., Alloway, B.J. and Thornton, I., 1975. An effect of soil ingestion on the utilisation of dietary copper by sheep. J. agric. Sci., Camb. 84 (2) 249-254.

Suttle, N.F. and Price, J. 1976. The potential toxicity of copper-rich animal excreta to sheep. Animal Production 23 (2) 233-242.

Unwin, R.J., 1980. Copper in pig slurry: some effects and consequences of spreading on grassland. In: Inorganic Pollution and Agriculture. MAFF Reference Book 326. HMSO.

Van Rhee, J.A. 1975. Copper contamination effects on earthworms by disposal of pig waste in pastures. Prog. in Soil Zoology Proc 5th Int. Coll. on Soil Zoology. Prague September 1972.

Williams, E.G. 1980. The Macaulay Institute for Soil Research Annual Report No 49 1978-1979 pp 85.

Williams, J.H., 1975. Use of sewage sludge on agricultural land and the effects of metals on crops. Journal of the Institute of Water Pollution Control, No 6. 635-644.

APPENDIX I

Using census data for 1975 (MAFF, 1978) and assuming a copper
loading of 80 g per pig place per year (i.e. to allow for
breeding stock) we can produce an assessment of the potential
copper distribution on land of pig holdings.

Stocking rate pigs/ha crops and grass	Rate of copper kg/ha per year	Number of holdings	Proportion of total herd accounted for %	Total area crops and grass ha	Proportion of total crops and grass %
> 50	4	2 965	29	12 500	0.13
> 35	2.8	3 467	33	19 700	0.21
> 25	2.0	4 268	38	30 000	0.31
Overall 3.8	0.3	33 291	100	1.8 million	19

* A similar calculation may be made annually as each year's census
information becomes available.

COPPER TOXICITY TO CROPS RESULTING FROM LAND
APPLICATION OF SEWAGE SLUDGE

M.D. Webber*, Y.K. Soon**, T.E. Bates** and A.U. Haq**

* Wastewater Technology Centre, Environmental Protection
Service, Department of the Environment, Burlington,
Ontario, Canada L7R 4A6.

** Department of Land Resource Science, University of Guelph,
Ontario, Canada N1G 2W1.

ABSTRACT

    *Sewage sludge from the town of Fergus in Ontario exhibited moderate
to severe toxicity to crops grown in lysimeter and greenhouse experiments.
The Fergus sludge contained large amounts of Cu, Zn and Cr and increased
the concentrations of these metals, particularly of Cu and Zn, in plant
materials. Other sludges which exhibited no toxicity produced similar
increases in the Zn and Cr concentrations of plant materials. These
findings indicated that the Fergus sludge toxicity to crops probably was
due to Cu. A large build up of Cu in the roots of annual ryegrass grown
on soil treated with Fergus sludge was further evidence for Cu toxicity.*

INTRODUCTION

Land application of sewage sludge is currently receiving considerable attention in Ontario. It is frequently the least expensive means of disposal and supplies valuable organic matter and nutrients to soils. Unfortunately, many sewage sludges contain heavy metals from industrial sources in addition to those present in human wastes. The dangers of toxicity to plants, animals and humans resulting from the indiscriminate use of sewage sludge containing large amounts of heavy metals have been discussed by several workers (Bates, 1972; Leeper, 1972; Page, 1974 and Webber, 1972).

One of the objectives of lysimeter experiments at Burlington and a greenhouse experiment at Guelph was to determine the toxicity to plants of Ontario sewage sludges containing large amounts of heavy metals. Heavy rates of sludge were applied to soils and several different crops were grown. In general, toxicity was observed only where sludge from the town of Fergus was applied. The Fergus sludge contained very large amounts of zinc, copper and chromium. The purpose of this paper is to present the evidence for copper toxicity to crops resulting from land application of Fergus sludge.

EXPERIMENTAL

Lysimeter Experiments - Burlington

One experiment with orchard grass (*Dactylis glomerata* L) cultivar Frode grown during 1978 and 1979 employed fluid sewage sludges. A second experiment with Swiss chard (*Beta vulgaris* var. *cicla*) cultivar Fordhook Giant grown during 1979 employed air-dried sewage sludges.

The fluid sludge experiment employed Brady fine sand (Table 1) in lysimeters which were 30.5 cm diameter x 183 cm deep. Sludges from North Toronto, Fergus and Owen Sound

(Table 2) were applied to the soil surface at rates of 100, 200 and 300 kg/ha of nitrogen (N) and Guelph sludge was applied at rates of 300 and 600 kg/ha N. Sludge was applied four times during 1978 and three times during 1979, at the beginning of the growing season and following all but the final harvest of orchard grass each year.

TABLE 1

SOIL PROPERTIES (0 - 15 CM DEPTH)

Soil	pH 0.01 M CaCl$_2$	Organic matter %	CEC meq/100g
Brady fine sand	4.4	< 1.0	0.6
Grimsby sandy loam	4.6	1.0	7.6
Conestoga loam	7.2	4.9	14.2

The air-dried sludge experiment employed Brady fine sand and Conestoga loam (Table 1) in lysimeters that were 61 cm x 61 cm x 70 cm deep. Sludges from North Toronto, Fergus, Owen Sound and Hamilton and sludge compost from Windsor (Table 2) were incorporated into the 0 - 15 cm layer of soil at two application rates. Rate 1 was 47 metric tons of sludge solids per hectare (47 tonnes/ha) applied in the fall of 1978 and at the beginning of the 1979 growing season. Rate 2 ranged from 200 to 500 tonnes/ha of sludge solids and was applied once only in the fall of 1978. It supplied 6 500 kg/ha N which was equivalent to the largest N loading to a previous lysimeter experiment with air-dried sludges.

Commercial fertiliser was applied to the control treatments of both lysimeter experiments to supply N, phosphorus (P) and potassium (K) for optimum crop growth. In addition, muriate of potash was applied to the sludge and compost treatments of both experiments to maintain adequate K for optimum crop growth.

Greenhouse Experiment - Guelph

Annual ryegrass (*Lolium multiflorum* L) was grown on soil

TABLE 2

TYPICAL ANALYSES FOR THE SLUDGES

Source	pH	Solids	N	P	Cd	Zn	Cu	Ni	Pb	Cr
			g/1				mg/1			
Fergus	7.2	46	2.50	2.00	0.35	1 238	1 048	11	11	540
Guelph	7.4	36	2.01	1.50	5.01	680	95	3	71	120
Sarnia	7.2	52	1.65	0.80	3.56	737	26	2	260	6
Midland	6.7	38	0.70	2.30	0.26	247	24	92	17	133
Aurora	7.3	36	1.70	1.50	0.11	42	15	2	11	360
North Toronto	7.3	39	1.75	0.86	0.89	72	62	1	66	32
Owen Sound	–	263	3.10	0.94	1.70	538	174	2	210	171
			mg/g				µg/g			
Hamilton	–	946	26.1	22.3	27.0	5 510	1 080	406	1 470	2 160
Windsor (Sludge compost)	–	959	11.7	11.7	13.9	1 610	376	109	341	408

treated with fluid sludges from North Toronto, Fergus, Guelph,
Sarnia, Midland and Aurora (Table 2). Soil from the plough
layer of a Grimsby sandy loam (Table 1) was adjusted to pH 7
with calcium carbonate and placed in plastic lined 30 cm
diameter pots. Fourteen crops of ryegrass were grown and
sludge at rates supplying 200, 800 and 1 600 kg/ha N was mixed
throughout the soil prior to seeding each crop. Commercial
fertiliser applied to a control treatment supplied N, P and
magnesium for optimum crop growth and muriate of potash applied
to all treatments supplied K for optimum crop growth. Following every second crop, the soil was removed from the pots and
leached with water to prevent soluble salt build up.

Analytical

Following harvest the plant materials were dried at $70^{\circ}C$
for 24 hours, weighed, ground in a stainless steel Wiley Mill
and stored in plastic containers. Samples for analysis were
ashed at $450^{\circ}C$ for 16 hours and the ash extracted with hot
aqua regia. Similarily, sludge samples for analysis were
dried, ground and extracted with aqua regia. Heavy metals in
the extracts were measured by atomic absorption spectrometry.

RESULTS

Lysimeter Experiment: Fluid Sludge

Orchard grass grew satisfactorily during the first year
(1978) of the experiment with fluid sludges. Yields were
similar for the different sludges and increased with increasing
rates of sludge application, particularly at harvests 3 and 4.
Reduced yields observed for all sludges at the 100 kg/ha N rate
probably resulted from N deficiency because the control yield
was similar to those produced by the higher sludge rates.
Harvest 3 yield data for the control and Fergus sludge treatments (Table 3) were typical of yield data collected during
1978. Copper (Cu) and Zinc (Zn) concentrations in grass grown
on the Fergus sludge treatments were larger than for the
control treatment but there was no evidence of toxicity.

TABLE 3

YIELD AND METAL CONTENT OF ORCHARD GRASS AS AFFECTED BY REPEATED
APPLICATIONS OF FLUID FERGUS SLUDGE TO BRADY FINE SAND IN LYSIMETERS

Harvest No.	Control	——————— Sludge ———————		
		N (kg/ha)		
	100	100	200	300
		Dry matter (tonne/ha)		
3*	3.24	2.71	3.12	3.05
5	1.41	0.98	0.69	0.17
7	2.15	0.95	1.70	0.66
		Plant Cu (µg/g)		
3	8.5	24.8	18.2	21.2
5	8.1	25.2	33.3	57.2
7	10.6	19.6	46.7	33.9
		Total Cu added (kg/ha)		
	0	298	596	895
		Plant Zn (µg/g)		
3	42	69	67	80
5	28	295	380	597
7	49	271	362	637
		Total Zn added (kg/ha)		
	0	346	692	1 038
		(Plant Cr (µg/g)		
3	1.3	4.1	5.7	4.7
5	1.5	1.6	4.0	5.7
7	2.6	6.5	12.7	10.9
		Total Cr added (kg/ha)		
	0	131	262	393

* The harvest 3 plant material was grown during 1978 and the harvests 5
and 7 plant materials were grown during 1979.

Although orchard grass is a perennial, much of the crop
died during the winter of 1978 - 79 and it was planted again in
the spring of 1979. During seedbed preparation, sludge applied
to the soil surface in 1978 was mixed throughout the 0 - 15 cm
layer.

Immediately following germination, in 1979, orchard grass growing on the Fergus sludge treatments exhibited toxicity symptoms. The foliage was yellow-green and growth and tillering were reduced. Moreover, the degree of toxicity increased with increasing sludge application rates. As the plants became established, their colour improved but growth remained severely reduced throughout 1979 as indicated by the yield data for harvests 5 and 7 (Table 3). By contrast, orchard grass growing on the North Toronto, Guelph and Owen Sound sludge treatments exhibited normal colour and growth increased with increasing rates of sludge application.

The Fergus sludge treatments added large amounts of Cu, Zn and chromium (Cr) to the soil and increased the concentrations of these metals in orchard grass (Table 3). The increases were greater in 1979 than in 1978. The concentrations of cadmium (Cd), nickel (Ni) and lead (Pb) were not increased indicating that toxicity was due to Cu, Zn and/or Cr. However, Guelph sludge applied at the 600 kg/ha rate added 458, 1 153 and 454 kg/ha Cu, Zn and Cr, to the soil and resulted in orchard grass with concentrations of 21, 192 and 11.5 µg/g of these metals, respectively. Since Guelph and Fergus sludges produced similar large Cr concentrations in orchard grass and Guelph sludge did not cause toxicity it is unlikely that the Fergus sludge toxicity was due to Cr.

Lysimeter Experiment: Air-Dried Sludge

Air-dried sludges added to Conestoga loam caused little change of soil pH, except for the Fergus sludge. Application rate 2 of Fergus sludge reduced the pH from 7.2 to 6.6 (Table 4) whereas values observed for the Hamilton, North Toronto, and Owen Sound sludge treatments and the Windsor sludge compost treatment were $\geq$ 6.9. In general, Swiss chard grew satisfactorily on the sludge treated soil. However, application rate 2 of Fergus sludge exhibited yield depression at harvest 1 which disappeared by harvest 3. The harvest 3 yields for rate 2 of all sludges were much larger than yields for rate 1, probably because the nitrogen supply was larger.

TABLE 4

SOIL pH, YIELD AND METAL CONTENT OF SWISS CHARD AS AFFECTED BY APPLYING
AIR-DRIED FERGUS SLUDGE TO TWO SOILS IN LYSIMETERS

Harvest No.	Contestoga loam			Brady fine sand		
	Sludge rate			Sludge rate		
	0	1	2	0	1	2
Soil pH (0.01M CaCl$_2$)						
Autumn	7.2	6.9	6.6	4.4	5.1	5.6
Dry Matter (tonne/ha)						
1	1.59	3.28	0.93	0.56	0.42	0.00
3	0.37	1.26	2.67	0.39	0.85	0.00
Plant Cu (μg/g)						
1	12	20	28	14	26	--
3	15	62	57	25	37	--
Total Cu added (kg/ha)						
	0	2 250	5 350	0	2 250	5 350
Plant Zn (μg/g)						
1	52	242	455	206	926	--
3	73	693	700	471	702	--
Total Zn added (kg/ha)						
	0	2 800	6 650	0	2 800	6 650
Plant Cr (μg/g)						
1	3.4	3.4	4.0	4.4	4.9	--
3	3.3	3.4	5.6	5.8	4.5	--
Total Cr added (kg/ha)						
	0	1 000	2 400	0	1 000	2 400

Sludge rate 1 was 47 tonnes/ha of sludge solids applied in the autumn of
1978 and again in the spring of 1979. Sludge rate 2 was 224 tonnes/ha of
sludge solids applied in the autumn of 1978.

Air-dried sludges increased the pH of Brady fine sand in the order Fergus, Hamilton, North Toronto, Windsor and Owen Sound. Values for the application rate 1 treatments were 5.1, 5.7, 5.8, 6.2 and 6.4, respectively, and were 0.5 to 0.6 units smaller than for the rate 2 treatments. Fergus sludge exhibited severe toxicity to Swiss chard grown on the Brady soil. Application rate 1 of Fergus sludge reduced the yield at harvest 1 and rate 2 killed the crop (Table 4). Application rate 2 of Windsor sludge compost was the only other treatment that exhibited toxicity. With this treatment yield depression was observed at harvest 1 but not at harvest 3. The harvest 3 yields for application rate 2 of all sludges except Fergus were larger than yields for rate 1.

The Fergus sludge treatments added large amounts of Cu, Zn and Cr to the soils and increased the Cu and Zn concentrations in Swiss chard plant material (Table 4). The Cr, Cd, Ni and Pb concentrations in plant material exhibited little if any increase indicating that Fergus sludge toxicity to Swiss chard was due to Cu and/or Zn. The other sludge treatments, except for Hamilton, produced smaller Cu and Zn concentrations in the plant material than the Fergus sludge treatments. The Hamilton sludge treatments produced similar large Zn concentrations. Since Hamilton sludge exhibited no toxicity to Swiss chard it is probable that the Fergus sludge toxicity was due to Cu.

Greenhouse Experiment: Fluid Sludge

Three crops of annual ryegrass per year were grown in the greenhouse during January to September. There was considerable variation in yield between seasons but growth was generally satisfactory and yields increased with increasing application rates of the Guelph, Sarnia, North Toronto, Midland and Aurora sludges. By contrast, Fergus sludge was toxic to ryegrass and yields decreased with increasing application rates of this sludge (Table 5). There was no yield from the 1 600 kg/ha N treatment of crop 14 and only a small yield of crop 13

TABLE 5

YIELD AND METAL CONTENT OF ANNUAL RYEGRASS AS AFFECTED BY REPEATED
APPLICATIONS OF FLUID FERGUS SLUDGE TO GRIMSBY SANDY LOAM IN THE GREEN-
HOUSE

Tissue	Crop No.	Control	Sludge		
		N (kg/ha)			
		200	200	800	1 600
		Dry matter (tonne/ha)			
Shoot	1	1.35	1.43	0.95	0.93
Shoot	5	0.80	0.65	0.61	0.23
Shoot	9	1.68	0.84	1.34	0.79
Shoot	14	0.64	1.01	0.84	0.00
		Plant Cu (μg/g)			
Shoot	1	11	16	21	21
Shoot	5	14	15	19	26
Shoot	9	13	18	34	34
Shoot	14	15	29	41	(49)
Root	14	88	270	1 108	--
		Plant Zn (μg/g)			
Shoot	1	43	35	46	43
Shoot	5	33	53	211	119
Shoot	9	36	99	170	224
Shoot	14	12	92	216	(559)
Root	14	42	100	271	--
		Plant Cr (μg/g)			
Shoot	1	1.5	3.3	2.3	2.3
Shoot	5	1.3	0.9	0.9	1.6
Shoot	9	1.2	1.8	2.9	2.5
Shoot	14	1.0	1.7	1.6	--
Root	14	2.0	9.0	25	--

Values in parentheses are for the 13th crop of ryegrass. There was no
yield from the 1 600 kg/ha N treatment of crop 14.

indicating that the degree of toxicity probably also increased as the number of sludge applications increased. The toxicity symptoms observed for ryegrass were similar to those described previously for orchard grass. There was no satisfactory explanation for the small yield from the control treatment of crop 14.

The Cu and Zn concentrations in ryegrass shoots grown on the Fergus sludge treatments were larger than for the control treatment and increased with the number and rate of sludge applications (Table 5). The Cr concentration in ryegrass shoots exhibited little, if any, variation with the number and rate of sludge applications. The Cu, Zn and Cr concentrations in crop 14 roots increased with increasing rates of sludge application.

Fourteen applications of Fergus sludge applied at the 1 600 kg/ha N rate added very large amounts of Cu, Zn and Cr to the soil (Table 6). However the same rate of Sarnia sludge added a larger amount of Zn and of Aurora sludge added a larger amount of Cr than the Fergus sludge. A comparison of metal concentrations in crop 13 and 14 shoots showed some variation between harvests but indicated that the highest Cu and Zn concentrations were exhibited by shoots grown on the Fergus and Sarnia sludge treatments, respectively. There was little variation of Cr concentration between sludge treatments. The Zn concentration in roots grown on the Sarnia sludge treatment was 184 µg/g and the Cr concentration in roots grown on the Aurora sludge treatment was 17 µg/g.

Since neither Sarnia nor Aurora sludge was toxic to annual ryegrass it is unlikely that Zn and Cr contributed to Fergus sludge toxicity. It is probable that the toxicity was due to Cu as indicated by high levels of this metal in the shoots and a massive build up in roots.

TABLE 6

METAL ADDITIONS TO SOIL IN 14 APPLICATIONS OF SIX FLUID SLUDGES APPLIED AT THE 1 600 kg/ha N RATE AND METAL CONTENT OF ANNUAL RYEGRASS AS AFFECTED BY THAT RATE OF APPLICATION

Sludge	Cu		Zn		Cr	
	\multicolumn{6}{c}{Added to soil (kg/ha)}					
Fergus	7 300		8 400		3 800	
Guelph	1 600		5 700		1 700	
Sarnia	620		12 700		140	
North Toronto	740		1 300		320	
Midland	860		4 400		2 600	
Aurora	280		550		5 500	
	\multicolumn{6}{c}{In ryegrass shoot (µg/g)}					
	\multicolumn{6}{c}{Crop No.}					
	13	14	13	14	13	14
Fergus	49	--	559	--	--	--
Guelph	33	39	198	207	1.7	1.2
Sarnia	34	27	648	492	3.0	1.8
North Toronto	28	27	190	117	2.4	1.7
Midland	29	24	192	121	1.6	1.5
Aurora	19	16	79	53	1.6	1.4

DISCUSSION

Reduced crop yields and chlorosis observed in the lysimeter and greenhouse studies following applications of Fergus sludge to soil were typical of heavy metal toxicity symptoms reported by Chaney and Giordano (1977). These authors suggested that chlorosis may result from excesses of several different metals and that it appears to be due to a direct or indirect interaction with foliar iron.

Small to moderate increases in the Cr and Cu contents and large increases in the Zn contents of crops following Fergus sludge treatment were in agreement with reports for other sludges containing large amounts of these metals. A task force to appraise the potential hazards of heavy metals in sewage sludge to plants and animals reported in 1976 (EPA 430/9-76-013) and a resume of their findings for Cr, Zn and Cu follows:

- Most crops absorb relatively little Cr, but some species can contain levels up to 10 μg/g.

- Zinc is taken up by plants as Zn^{2+} and, in excessive quantities, can be toxic. However, few records of toxic effects of Zn are available in the literature.

- Copper toxicity may develop in plants from application of sewage sludge if the Cu concentration in sludge is relatively high. Copper concentrations in plants normally do not build up to high levels when toxicity occurs. Most of the excess Cu accumulates in the plant roots.

Copper and Zn concentrations exceeding 20 and 200 μg/g, respectively, in the orchard grass, annual ryegrass and Swiss chard shoots were in the toxic ranges as reported by Allaway (1968) and Beckett et al. (1979). However, findings reported here and by the task force implicate Cu as the most likely cause of Fergus sludge toxicity.

Metal interactions were not considered in the present study, however, they can affect uptake and toxicity as reported by Cunningham et al. (1975) and may have contributed to Fergus sludge toxicity.

REFERENCES

Allaway, W.H., 1968. Agronomic controls over the environmental cycling of
trace elements. Advan. Agron. 20, 235-274.

Bates, T.E., 1972. Land application of sewage sludge. COA Research
Report No. 1, Ottawa.

Beckett, P.H.T., Davis, R.D. and Brindley, P., 1979. The disposal of
sewage sludge onto farmland: The scope of the problem of toxic
elements. Wat. Pollut. Control 78, 419-445.

Chaney, R.L. and Giordano, P.M., 1977. Microelements as related to plant
deficiencies and toxicities. In L.F. Elliot and F.J.
Stevenson (Eds.) Soils For Management of Organic Wastes. Amer. Soc.
Agron. Inc., Madison, Wisc., USA. pp 233-279.

Cunningham, J.D., Ryan, J.A. and Kenney, D.R., 1975. Phytotoxicity in and
metal uptake from soil treated with metal-amended sewage sludge.
J. Environ. Qual. 4, 455-460.

EPA 430/9-76-013, 1976. Application of Sewage Sludge to Cropland:
Appraisal of potential hazards of the heavy metals to plants and
animals. US Environmental Protection Agency, Washington, DC,
20460. 63 pp.

Leeper, G.W., 1972. Reactions of heavy metals with soil with special
regard to their application in sewage wastes. US Dept. of Army.
Corps of Engineers. Contract No. DACW 73-73-C-0026. 70 pp.

Page, A.L., 1974. Fate and effects of trace elements in sewage sludge
when applied to agricultural lands. A literature review study.
U.S. Environmental Protection Agency, Report No. EPA 670/2-74-005,
Washington, DC, 20460. 108 pp.

Webber, J., 1972. Effects of toxic metals in sewage on crops. Wat.
Pollut. Control 71, 404-413.

DISCUSSION

R. Braude *(UK)*

First, may I ask on what you base your conclusion that the various effects you mentioned are due to copper. Secondly, in your control plot, as far as the roots were concerned, you had a very big increase. What do you think that was due to?

M.D. Webber *(Canada)*

Well, it was not due to sludge because there was no sludge added to the control treatment. There are problems with root measurements; part of the increase may have been related to some small amounts of contamination on the root material, although every effort was made to clean the roots as well as possible. The only other suggestion I can offer is that normally the root of ryegrass would have a higher copper content than the shoot.

To come back to your other question concerning the basis for implicating copper, as I mentioned, we used five or six sludges although I only showed data for the Fergus sludge. All the sludges were heavily contaminated with metals, although the others were not so heavily contaminated as the Fergus sludge. In the case of the Swiss chard, a Hamilton sludge treatment gave us very similar high zinc contents in the plant material but no apparent toxicity and no decrease in yield. The differences in the observations from the Hamilton sludge treatment and the Fergus sludge treatment were the higher contents of copper in the plant material produced on the Fergus sludge. In the case of ryegrass, a Sarnia sludge treatment produced ryegrass with a higher zinc content than the Fergus sludge did. An Aurora sludge treatment supplied more chromium to soil but neither the Aurora nor Sarnia sludge treatments produced any apparent toxicity problems. The evidence is circumstantial but it does point towards copper being a problem.

P.H.T. Beckett *(UK)*

I think there is fairly definite evidence of copper toxicity.

It is a useful rule of thumb that if the copper content of young vegetative tissue exceeds 20 ppm, you are on the verge of toxicity The figures in Table 4 show that the 1 and 2 sludge rates are well in excess of 20 ppm. What is curious is that the control of the Brady Fine Sand, on this criterion with no sludge, shows both copper and zinc toxicity. The Brady Fine Sand seems to have started in a bad way.

M.D. Webber

Obviously it did and perhaps I can be criticised for not having limed the Brady Fine Sand before doing the experiment. The pH of that soil was very low. The soil was chosen because it was acid, because it was indeed very sandy. It contains less than 1% of clay material. In fact, the crops grown on that soil were suffering from the effects of soil acidity in addition to an effect of adding sludges containing high metal contents.

R.J. Unwin *(UK)*

Could you tell us what soil you were using in the greenhouse experiment?

M.D. Webber

It is a Grimsby sandy loam; the pH of that soil was adjusted to 7. As it exists in the field it is somewhat acid.

R.J. Unwin

You have shown us yield figures for the one sludge for the ryegrass. Did you have yield effects with your other sludges in the greenhouse? There is quite a variation in overall levels of metal addition and I was wondering where any yield effects may have appeared.

M.D. Webber

The yield effects with the other sludges, if anything, were positive, compared to the control. The Fergus sludge was the only sludge which gave us any deleterious effect on yield. I might add that the town of Fergus owns some property beside the town and they have been loading this sludge onto their soil heavily for at least 15 years. This summer I went to the site where the sludge has been applied; it was growing a mixture of wheat and barley and the crop appeared to be growing well. We have taken some soil and crop samples and we are going to follow up this situation.

The authorities in Fergus are under pressure to clean up their effluent. There are two industries in Fergus, one of them involved in making metal alloys, brasses and so on. I understand that they have hired an environmental engineer whose job it will be to improve the quality of their effluent.

M. Lineres *(France)*

How were you able to measure the metal contents of roots without risk of pollution by the sludge?

M.D. Webber

We measured as carefully as we could. We recognise that in doing root analysis, there is a potential for contamination from soil and from sludge. The roots were recovered from the soil; in this case a sample of root was recovered because one recognises that with grass one does not recover the whole root. The samples were washed as carefully as possible, dried and analysed.

R.D. Davis *(UK)*

Is there not a possibility that in all this washing you are washing away some copper out of the roots?

M.D. Webber

Yes, perhaps the copper contents ought to have been higher.

R.D. Davis

So it may be that roots are not a very good diagnostic organ for various reasons?

M.D. Webber

Yes, it may be so. In this situation, root analysis is a last resort. The fact that you do observe such high levels of copper in the roots where there is obviously a problem gives, I think, further evidence that probably copper is the main problem.

G.A. Fleming (Ireland)

You quoted a figure of 49 ppm of copper indicating phytotoxicity. Dr. Beckett gave a figure of about half that, 20 ppm. I think there are a few things we must bear in mind here. You are speaking of material grown in pots. If the same material had been grown in the field, I wonder whether a figure of 49 ppm would have been attained. Also, I would say that it is quite difficult to get a copper content higher than 20 ppm in ryegrass in the field, even with moderate applications of things like copper sulphate. I wonder if anyone would care to hazard a guess at what level of copper application phytotoxicity might begin under field conditions?

M.D. Webber

I do not think I suggested that 49 ppm was the level at which toxicity began to occur, or that that was a critical level. That is simply the level which we observed in this plant material which was obviously under stress - the yield was considerably reduced.

J.K.R. Gasser

In relation to the root contamination with soil, Mitchell at the Macaulay Institute used to use iron content as an indicator of contamination of plant samples for trace element work. I wonder whether you have done any iron analyses on these samples. If so, it would be quite interesting to see whether they give any indication of soil contamination, and if so, how much.

M.D. Webber

I cannot answer your question but it is certainly worth taking note of.

G.A Fleming

May I suggest that instead of measuring iron, you measure titanium.

COPPER TOXICITY IN VITICULTURAL SOILS

J. Delas

Institut National de la Recherche Agronomique,
Station d'agronomie - Centre de Recherches de Bordeaux,
33140 Pont de la Maye
France.

Copper salts have been widely used as fungicides in the French vineyards since 1885, and they were the only weapon which could be used against the downy mildew of grapevine (*Plasmopara viticola*) until organic or organo-metallic substitution compounds appeared seventy-five years later. Thus considerable amounts of metal have accumulated in soils.

Due to the copper toxicity for fungi, the possible effects of this metal on soil fertility were studied at an early date (Girard, 1895; Maquenne and Demoussy, 1919): reassuring results were obtained as soil acidity, later to be recognized as a main factor in toxicity, had been disregarded.

Possible serious effects of copper accumulation were only observed much later, after 1950, after a great deal of damage had been noticed in French vineyards (Anne and Dupuis, 1953; Drouineau and Mazoyer, 1953; Depardon and Buron, 1955; Delas et al., 1959). Several studies were then conducted on copper toxicity, as shown in our bibliography (Delas, 1963).

In the present paper, we will deal mainly with results obtained in Bordeaux which are in agreement with data collected in France or foreign countries.

1) Occurrence of toxicity

Investigations were undertaken (Delas et al., 1959; Delas et al., 1960; Delas, 1967; Delas and Dartigues, 1970) after an increased occurrence of a particular type of damage had been observed in the Bordeaux vine. The trouble appeared after the uprooting of an old vine and renewed culture (planting of vine or seeding of annual plants). These anomalies consist in a high mortality of young vines, a non-existant or irregular sprouting of annual plants and a considerably reduced growth. Foliar symptoms are not specific. An atrophy of the root system can be observed in all cases.

The study conducted in 1958 and 1959 reveals that such anomalies occur in soils presenting the following common characteristics: high copper content (up to 1 000 ppm of total copper and 200 ppm of exchangeable copper (ammonium acetate)), and acid pH. Our pot and field experiments confirmed the phytotoxicity of such copper concentrations in acid soils, as already noted by several authors.

Characteristic features of the Atlantic climate and the soil evolution resulting therefrom contribute to the serious problem of copper toxicity in the Bordeaux vineyards: due to spring rain, numerous fungicide treatments are required to protect high quality vines. From 1885 until recently, these treatments were performed with 'bouillie bordelaise' (a mixture of lime and copper sulphate), thereby supplying 15 to 50 kg of copper/ha/year. Moreover, the Atlantic climate indirectly affects the increase in copper toxicity: soil acidification is caused by leaching phenomena and even reinforced by sulphur applied to fight against the powdery mildew of grapevine *(Uncinula Necator)* (Pochon et al., 1962; Pochon and Chalvignac, 1965).

2) Conditions of toxicity occurrence

Two conditions are required for toxicity occurrence: the soil must be acid and it must contain a sufficient amount of available copper. The more acid the pH, the higher the toxicity. Another factor is the cation exchange capacity (CEC). For the same copper concentration, an adverse effect is more likely to occur with a low CEC. The toxic level for a water pH less than 6 varies between 25 ppm of exchangeable copper (ammonium acetate) in sandy soil (Delas et al., 1960) and 100 ppm in clay soil (Drouineau and Mazoyer, 1962). This toxic level also varies according to plants, leguminosae appearing to be more sensitive.

3) Behaviour of copper accumulated in soil

In viticultural soils, which are usually very poor in organic matter, copper migrates little in depth and concentrates in the top layer (Table 1).

TABLE 1

COPPER FIXATION IN SURFACE (DELAS, 1959)

Depth (cm)	Exchangeable copper (ammonium acetate)
0 - 25	170
25 - 50	25
50 - 75	traces

The old vine resists due to its root system which lies underneath the copper-enriched zone: troubles occur only after uprooting, when new plants have been set in the superficial layer containing the accumulated metal.

Thus, except for neutral or calcareous soils in which copper is precipitated in the form of poorly soluble carbonates or hydroxydes, copper is available for plants as well as energetically fixed in the surface by clay and humic colloids.

Using a method of isotopic dilution with ^{64}Cu, we showed that a large fraction of copper which had been accumulating in viticultural soils for seventy-five years remains in the soil in diffusible form, and that the isotopic 48 hours diluted fraction corresponds approximately to the exchangeable copper (ammonium acetate) (Table 2).

As the activity of copper ions increases continuously with the amounts retained in the soil, it is possible that the metal toxicity might also affect soils where it has not yet been identified.

TABLE 2

FORMS OF COPPER IN VITICULTURAL SOILS IN THE BORDEAUX AREA
(DELAS ET AL., 1960)

Copper concentration (ppm)	Reference of samples									
	1	2	3	4	5	6	7	8	9	10
Total copper	305	328	335	355	406	440	480	525	565	845
Isotopic 48 h diluted copper	97	132	104	109	152	289	186	154	121	168
Exchangeable copper (ammonium acetate)	64	86	78	75	130	200	170	139	136	186

While viticultural soils show a poor in-depth migration of copper, this is not the case for soils rich in organic matter, such as humoferruginous soils in which copper can migrate in depth in the form of organo-metallic compounds together with the most mobile humic compounds (Delas, 1967).

4) Treatment of copper toxicity

Effects of copper toxicity can be considerably reduced by applying lime to increase the soil pH to approximately 6 - 6.5. The effect of liming is illustrated by the results of a pot experiment which was conducted to study the sensitivity of copper to non-rooted vine cuttings (Table 3).

In acid soil, copper incorporation considerably reduced the growth of cuttings. After liming, the same copper application has no effect on the cuttings' growth.

In spite of the generally very satisfactory results of liming, some cases are noted when massive lime applications do not eliminate toxicity effects. For instance, an experiment conducted by the Compagnie d'Aménagement du Bas-Rhône et

TABLE 3

EFFECT OF COPPER APPLICATIONS AND LIME ON THE WEIGHT INCREASE OF VINE
CUTTINGS GROWN IN POTS (DELAS AND DARTIGUES, 1970)

| | (g/pot) | | | | | |
| | non limed soil (water pH 4.8) | | | limed soil (water pH = 6.1) | | |
	'initial weight	final weight	increase' %	'initial weight	final weight	increase' %
Control with-out copper	4.31	8.12	88.3	4.64	9.01	94.1
Application of 240 mg of copper per kg of earth	4.57	5.31	16.1	5.13	9.40	83.2

Languedoc indicated that applications of 4 t/ha of Dolomagnésie
on copper enriched soils do not produce yields as great as on
non enriched soils (Figure 1).

An additional method of reducing the effects of excess
copper consists of increasing the humic level of soils: the
metal activity is decreased by organic amendments which either
block copper in the form of insoluble organo-metallic compounds
or induce copper migration in the form of soluble compounds.

In the long term, the problem of copper accumulation in
viticultural soils could be solved by stopping all applications.
About 20 years ago, the development of organic or organo-
metallic fungicide compounds containing little or no copper
resulted in the lesser use of traditional copper sprays. However, products based on copper are still used nowadays (though
to a small extent) so that vineyards continue to be affected
by copper toxicity.

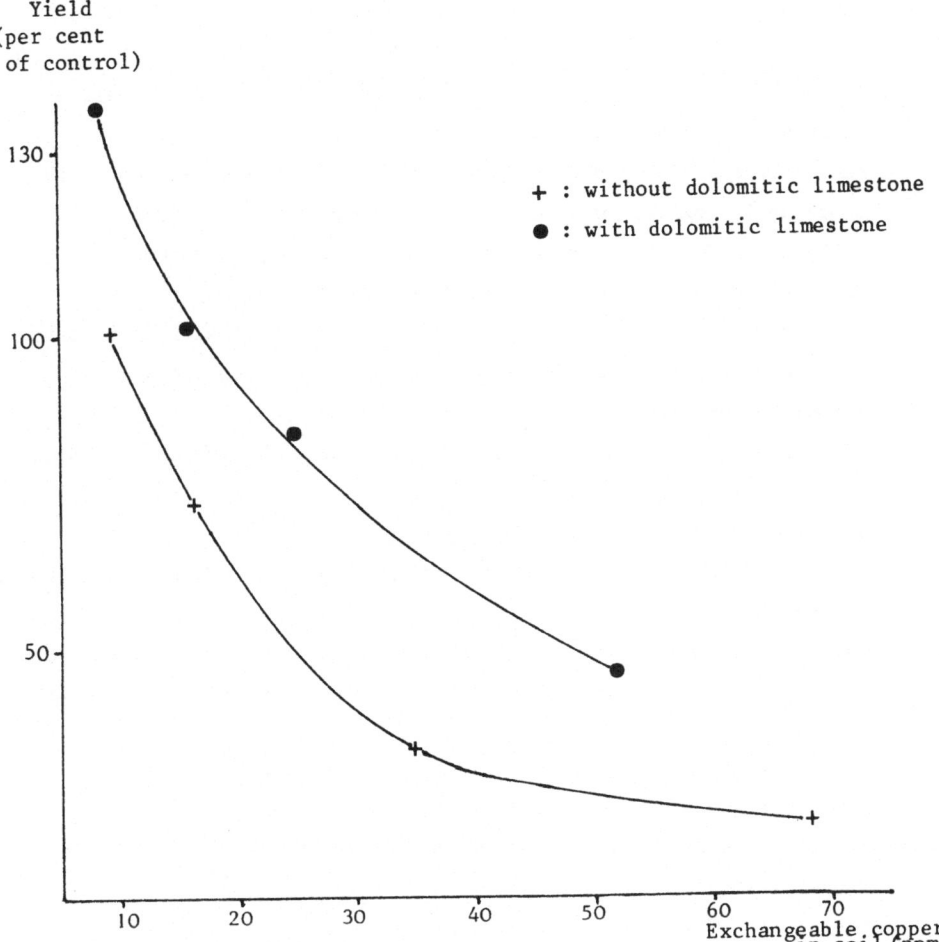

Fig. 1. Relationship between the weighted yield of 12 cultures and the concentration of exchangeable copper in the soil (data given by la Compagnie Nationale d'Aménagement du Bas-Rhône et Languedoc, 1966).

REFERENCES

Anne, P. and Dupuis, M., 1953. Toxicité du cuivre à l'égard de quelques plantes cultivées. C.R. Acad. Agric. 39, 58-60.

Compagnie Nationale d'Aménagement du Bas-Rhône et Languedoc (CNABRL), 1966. Etude de la toxicité d'apports de sulfate de cuivre effectués en 1962 et à differents niveaux, associés ou non à un amendement calcaire (dolomie). C.R. activité Champ d'essai d'Asport (îlot 8)

Delas, J., Delmas, J., Rives, M. and Baudel, C., 1959. Toxicité du cuivre dans les sols viticoles du Sud-Ouest atlantique. C.R. Acad. Agric. 45, 651-655.

Delas, J., Delmas, J. and Demias, C., 1960. Etude par dilution isotopique du cuivre incorporé dans les sols depuis 70 ans. C.R. Acad. Sci. 250, 3867-3869.

Delas, J., 1963. La toxicité du cuivre accumulé dans les sols. Agrochimica 7, 258-288.

Delas, J., 1967. Liaisons entre cuivre et matière organique dans un podzol développé sur sable des Landes et accidentellement enrichi en cuivre. Ann. agron., 18, 17-29.

Delas, J. and Dartigues, A., 1970. Exemple de problèmes régionaux. II. Le Sud-Ouest. Ann. agron., 21, 603-615.

Depardon, L. and Buron, R., 1955. Cultures agricoles sur défrichement de Vigne. Ann. agron., 6, 161-167.

Drouineau, G. and Mazoyer, R., 1953. Toxicité du cuivre et évolution des sols sous l'influence des antiparasitaires. C.R. Acad. Agric., 39, 390-392.

Girard, A., 1895. Sur l'accumulation dans le sol de composés cuivriques employés pour combattre les maladies parasitaires des plantes. C.R. Acad. Sci., 120, 1147-1152.

Maquenne, L. and Demoussy, E., 1919. Sur la richesse en cuivre des terres cultivées. C.R. Acad. Sci., 169, 937-942.

Pochon, J., Leborgne, L., Tardieux, P. and Falcou-Sigrand, J., 1962. Influence des doses élevées de cuivre et de soufre sur la microflore bactérienne des sols viticoles (Médoc). Ann. Inst. Pasteur, Paris, 103, 614-622.

Pochon, J. and Chalvignac, M.A., 1965. Oxydation biologique du soufre et dégradation des sols de vignoble. Agrochimica, 9, 155-161.

DISCUSSION

M.D. Webber *(Canada)*

Is it necessary to continue adding lime after the new crop of grapes has been established?

J. Delas *(France)*

Lime application before planting is necessary for an abundant vine crop. It is not necessary once the plants have become established. The roots are in the non-enriched layer of the soil.

G.A. Fleming *(Ireland)*

Has deep ploughing been tried in order to translocate the copper from the top 5 cm layer down to 50 or 60 cm?

J. Delas

The disadvantage of doing that is that you are then spreading the copper through the soil making it impossible for the roots to find layers with a low copper content.

G.A. Fleming

Yes, one can see the point that in time the older vines might become affected but I would have thought, in looking at the soils we saw yesterday, that it would be relatively easy, with modern machinery, really to bury it down to perhaps 1 m.

Th. M. Lexmond *(The Netherlands)*

Deep ploughing of copper affected soils can only be expected to be of any value when there is some buffer power in the subsoil, when you have subsoils that are acid and do not contain any organic matter it is no use to dilute the copper because it would not affect the copper activity in the soil solution.

IMPLICATIONS OF APPLYING COPPER-RICH PIG SLURRY TO GRASSLAND - EFFECTS ON PLANTS AND SOILS

D. McGrath

The Agricultural Institute, Johnstown Castle,
Wexford, Ireland.

ABSTRACT

In a series of experiments pig slurry was applied to swards which were subsequently grazed or conserved. Analyses for copper were performed on plant and soil samples. Pot experiments were conducted to assist in the interpretation and generalisation of results.

Increases in herbage copper, which could be attributed to contamination by slurry, varied from less than 4 µg/g to greater than 100 µg/g. Management factors, season, rainfall pattern subsequent to spreading and herbage composition influenced herbage copper levels.

Uptake of copper by herbage occurred to an insignificant extent in the field under circumstances where uptake of zinc was easily demonstrated. Factors influencing uptake in pots included herbage composition and soil type: slurry copper was no more toxic to germinating ryegrass seedlings than was mineral copper.

Recoveries from soil of copper applied to pig slurry were low: added copper appeared to be totally extractable with aqueous 0.5 M EDTA.

INTRODUCTION

Faecal slurry from pig fattening units currently contains copper at a concentration of ca 700 µg/g DM. Consequences for soil, plant and animal from the application and disposal on pasture of this copper-rich material have recently been considered (McGrath et al., 1980). Some of the relevant distribution pathways for copper (Figure 1) are discussed in more detail below.

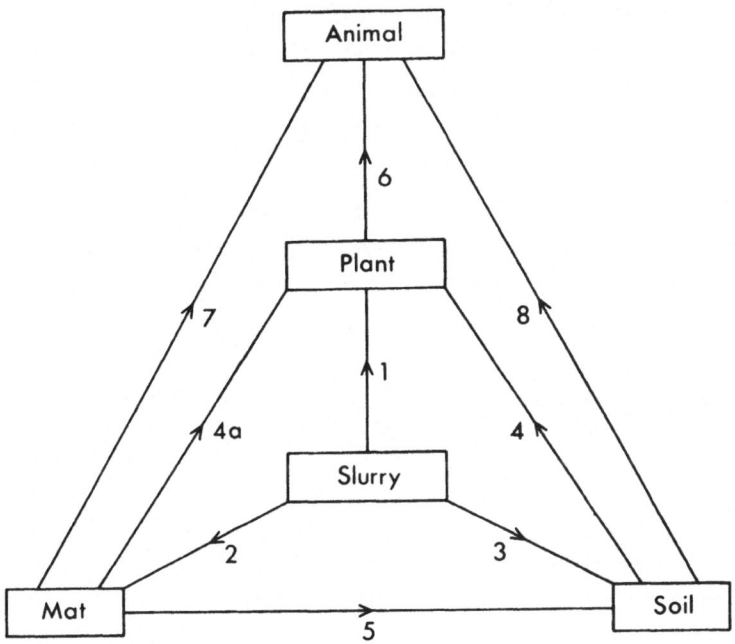

Fig. 1. Distribution of slurry-applied copper.

EXPERIMENTAL

Soils and experimental procedures have largely been described (McGrath et al., 1980).

RESULTS AND DISCUSSION

Copper in surface mat

Applications of pig slurry, especially when frequent and heavy, result in the formation of a mat on the soil surface (Dalgarno and Mills, 1975; McGrath et al., 1980). At each of 4 application sites, slurry had accumulated on the surface of plots after 3 years, forming a mat rich in organic matter. The average copper content of mat samples was 219 µg/g DM, whereas the average copper content of applied slurry had been 739 µg/g (Table 1). Even when a generous correction was made to reduce the ash content of samples from 50% to 10% (to represent the ash in washed slurry) copper content of mats averaged only 399 µg/g. Thus it appears that, while the mat persists on the soil surface, its copper content is being reduced by leaching. It is probable that mats do not persist for any length of time except where swards are cut exclusively. Even then they appear to be largely destroyed between the end of one season and the beginning of the next.

TABLE 1

COPPER IN SOIL (0 - 0.5 cm HORIZON) FROM CONTROL PLOTS AND IN MAT SAMPLES FROM TREATMENT (HIGHEST APPLICATION RATE) PLOTS

Site	Total slurry application (kg DM/ha)	Control Copper (µg/g)	Treatment	
			Ash[1] (%)	Copper (µg/g)
1	54.0	5.6	45.9	227 (412)[2]
2	63.1	7.5	53.2	216 (439)
3	64.0	14.0	43.5	215 (360)
4	51.5	10.6	44.8	220 (385)

[1] The ash contents of samples for control plots were within the range 85 - 92%

[2] Values corrected for ash (to 10%) are shown in parentheses

- 146 -

Copper in soil

As expected, most (Kornegay et al., 1976; Gracey et al., 1976) but not all (Batey et al., 1972) of the build-up of slurry copper occurred at or near the soil surface. Measurements for control plots and for plots which had received slurry applications totalling 1 000 m^3/ha over a period of 3 years are shown in Table 2.

TABLE 2

COPPER DISTRIBUTION IN SOIL HORIZONS 0 - 5, 5 - 10 AND 10 - 15 cm

Site	0 - 5 cm		5 - 10 cm		10 - 15 cm	
	Control	treatment	Control	treatment	Control	treatment
	Total copper content (µg/g) of soil					
1	5.5	28.0*[1]	4.5	7.5	3.5	4.5
2	7.5	27.5*	8.0	9.5*	8.5	9.0
3	15.0	47.4*	14.0	20.5*	14.5	14.0
4	13.5	51.5*	12.5	16.0*	14.8	16.5

[1] Significantly different from control at P = 0.05 using analysis of variance

For both 0 - 5 and 5 - 10 cm horizons, the average increase in EDTA (0.05 $\underline{M}$) - extractable copper was as great as, and indeed marginally exceeded, the increase in total copper. This would suggest a greater degree of mobility and plant-availability for slurry-derived copper than for native soil copper.

Zinc applied in pig slurry generally had soil distribution and solubility characteristics similar to copper.

It is of some concern that recoveries of copper applied in pig slurry were much lower than expected. They ranged from 38% for soil 1, a sand, to 69% for soil 4, a loam of high silt content: for estimation purposes it was assumed that 1 kg of copper distributed to a depth of 7.5 cm in soil raised its

copper content by 1.0 µg/g. Similar estimations using
published data indicate recoveries of 44% (Batey et al., 1972),
at least 89% (Gracey et al., 1976) and 131% (Kornegay et al.,
1976). Thus it appears that a loss of copper can occur in the
field and that it may depend on soil type. Removal of copper
on harvested herbage could account for no more than 2% of that
applied in slurry.

Copper uptake by herbage

The copper content of herbage may be raised in two ways
as a result of pig slurry application to grassland swards, viz
by (a) plant uptake and (b) contamination with slurry or soil-
copper particles.

Uptake of slurry-derived copper was studied in pot
experiments. The copper content of newly sown perennial ryegrass increased in similar manner with increasing soil
concentrations of either slurry or copper sulphate copper.
Herbage copper reached a maximum of 30 µg/g at a soil copper
concentration greater than 100 µg/g. Under field conditions
increased concentration in herbage was easily demonstrated for
zinc but not for copper where slurry had been repeatedly
applied to well established pasture in previous years. This
is reassuring in view of the high copper content of the top-
soil (Table 3).

TABLE 3

COPPER AND ZINC IN SOIL AND HERBAGE FROM SLURRY TREATMENT PLOTS (SITE 3)

	Copper (µg/g)		Zinc (µg/g)	
	Control	treatment	Control	treatment
Soil				
0 - 5 cm horizon	14.9	52.8	55.5	90.0
0 - 5 cm horizon	14.2	108.0	-	-
Herbage	9.4	10.0	36.5	50.4

In pot experiments copper uptake was found to vary considerably with herbage species. It was greatest for broad-leaved grasses like timothy *(Phleum pratense)* and cocksfoot *(Dactylis glomerata)* and least for perennial ryegrasses and clovers (Table 4). Similar uptake patterns were obtained when slurry was present as a mat on the soil surface with an established sward as when it was incorporated into soil and grass subsequently grown. Contrary to expectation (Fleming, 1973) copper contents of clovers were consistently lower than those of grasses.

TABLE 4

COPPER CONTENT OF FIRST CUT HERBAGE GROWING IN SOIL IN WHICH SLURRY AT 750 m³/ha HAD BEEN INCORPORATED TO A DEPTH OF 10 cm

Plant genotype	Control	Treatment
	Copper content (µg/g) of herbage	
Lolium perenne, S23	11.0	16.4***
Lolium perenne, Vigor	11.3	16.9***
Lolium perenne, Reveille	13.8	17.6***
Lolium multiflorum, Lemtal	11.8	15.6***
Festuca arundinacea, Alta	10.0	18.0***
Dactylis glomerata, Rano	16.8	25.8***
Phleum pratense, Climax	13.7	23.3***
Trifolium repens, Blanca	10.7	11.9 NS
Trifolium pratense, Hungaropoly	9.5	13.5***
Trifolium repens, Kentish wild white	12.0	11.9 NS

*** Significantly different from control at P = 0.001: NS not significant

At high concentrations in soil, copper is toxic to plants, especially at the seedling stage. However, much of the toxicity of pig slurry was associated with its high soluble salt concentration. It proved possible to reduce this greatly first by drying and then by carefully washing the solid pieces of slurry. Using this material, slurry copper at 200 µg/g in soil gave a

20% depression in the growth of ryegrass seedlings. Copper
sulphate gave the same effect at a concentration of 170 µg/g
copper in soil. Toxicity decreased with successive herbage
cuts. Grass established and grew vigorously in time, albeit
with reduced plant numbers, at a copper (sulphate) concentration as high as 390 µg/g in soil but failed completely at a
concentration of 780 µg/g. These toxicity experiments were
conducted in soil 3 (a loam). Toxicity is known to be
influenced by soil characteristics including pH and CEC (Baker,
1974) and could be expected to occur more easily in lighter
soils. Interestingly, very little difference in uptake of
copper by ryegrass from different soils was found although it
was marginally higher initially from soil 1 (a sand) than from
the other soils which were examined.

Copper contamination of herbage

Whereas measureable increases in plant copper, attributable to uptake, occur only after the addition to soil of large
amounts of slurry copper, large increases attributable to contamination can occur after relatively light applications to
swards. Plant contamination effects appeared to be more
pronounced under grazing than under cutting conditions. Thus
when slurry at 20 m^3/ha was applied 3 times annually to
paddocks being grazed by sheep, average copper content of
herbage, sampled at intervals of 14 days throughout the grazing
season, was considerably elevated (Table 5).

TABLE 5

AVERAGE COPPER CONTENT OF HERBAGE SAMPLED ON 2 GRAZING SITES IN 1977 AND
1978

Year	Site 3		Site 1a	
	Control	Slurry	Control	Slurry
Copper content of herbage (µg/g)				
1977	10.0	31.0	5.9	38.6
1978	10.6	42.1	4.9	33.1

On swards which were cut, copper of harvested herbage was shown to be clearly affected by management and other factors. The following were identified as being important in determining the degree of contamination: rate of application of slurry, sward height at time of application, rainfall pattern (timing and intensity) subsequent to application, and the interval between application and cutting.

In spreading trials, increases in herbage copper varied from less than 4 µg/g to greater than 100 µg/g for first cut silage following slurry application at a rate of 115 m^3/ha (Table 6).

TABLE 6

COPPER CONTENT (µg/g) OF FIRST CUT HERBAGE AT 4 SITES FROM PLANTS TREATED WITH SLURRY APPLIED AT RATES OF O AND 115 m^3/ha

Site	1976		1977		1978	
	Control	treatment	Control	treatment	Control	treatment
	Copper content (µg/g) of herbage					
1	3.5	19.7[1]	3.2	7.0	2.4	9.7
2	9.9	119.0	5.5	7.6	7.1	13.4
3	11.8	103.7	6.6	10.0	6.8	19.4
4	9.3	27.3	6.2	8.4	6.8	10.7

[1] All differences between treatment and relevant control were significantly different at P = 0.05

Smaller increases were measured following lower rates of application. Aftermaths harvested in October invariably had high copper contents with increases of 20 µg/g for herbage from high treatment plots and proportionate increases for lower rates of application. This was attributed to the process of contamination of herbage with soil particles which is accentuated at this time of the year (Healy, 1973). It is possible that uptake, which undergoes a seasonal increase (Fleming, 1973), may also have contributed.

The results of pot experiments indicated that degree of contamination varied between herbage species (Table 7). It was greatest for fine leaved grasses and clovers and least for coarse leaved grasses including tall fescue *(Festuca arundinacea)*.

TABLE 7

COPPER CONTENT OF HERBAGE AFTER TREATMENT WITH SLURRY AT 23 m^3/ha

Plant genotype	Control	Treatment
	Copper content (µg/g) of herbage	
Lolium perenne, S23	11.9	27.3***[1]
Lolium perenne, Vigor	13.5	36.5***
Lolium perenne, Reveille	14.0	23.7***
Lolium multiflorum, Lemtal	13.5	25.3***
Festuca arundinacea, Alta	10.3	16.1*
Dactylis glomerata, Perdice	11.6	14.9 NS
Phleum pratense, Glasnevin	8.2	10.3 NS
Trifolium repens, Blanca	10.8	38.5***
Trifolium pratense, Hungaropoly	14.5	25.9***
Trifolium repens, Kentish wild white	11.6	22.3***

[1] Treatment was significantly different from control at P = 0.001 (***) or 0.05 (*): NS not significant

REFERENCES

Baker, D.E., 1974. Copper: soil, water and plant relationships.
Federation Proceedings, 33, 1188-1193.

Batey, T., Berryman, C. and Line, C., 1972. The disposal of copper-
enriched pig-manure slurry on grassland. J. Br. Grassland Soc.
27, 139-143.

Dalgarno, A.C. and Mills, C.F., 1975. Retention by sheep of copper from
aerobic digests of pig faecal slurry. J. agric. Sci. Camb. 85,
11-18.

Fleming, G.A., 1973. Mineral composition of herbage in 'Chemistry and
Biochemistry of Herbage.' Vol I. Ed. G.W. Butler and R.W. Bailey,
Academic Press, London and New York, 529-566.

Gracey, H.I., Stewart, T.A., Woodside, J.D. and Thompson, R.H., 1976.
The effect of disposing high rates of copper-rich pig slurry on
grassland on the health of grazing sheep. J. agric. Sci. Camb,
97, 617-623.

Healy, W.B., 1974. Nutritional aspects of soil ingestion by grazing
animals in 'Chemistry and Biochemistry of Herbage' Vol. 1. Ed.
G.W. Butler and R.W. Bailey, Academic Press, London and New York,
567-588.

Kornegay, E.T., Hedges, J.D., Martens, D.C. and Kramer, C.Y., 1976.
Effect on soil and plant mineral levels following applications of
manures of different copper contents. Plant and Soil, 45, 151-162.

McGrath, D., Poole, D.B.R., and Fleming, G.A., 1980. Hazards arising from
application to grassland of copper rich pig faecal slurry in
'Effluents from Livestock'. Ed. J.K.R. Gasser. Applied Science
Publishers Ltd., London 420-431.

COPPER ACCUMULATION IN BRITTANY SOILS THROUGH
ENRICHED PIG SLURRY; PHYTOTOXIC RISKS

M. Coppenet
Station d'agronomie de l'Institut National de la
Recherche Agronomique, 4, Rue de Stang Vihan,
29000 Quimper, France.

ABSTRACT

*In Brittany pig breeding is particularly intensive and the
'Département du Finistère' was the subject of our special attention.
Copper complementation of feeds for fattening-pigs was often 125 ppm Cu
five years ago but today the tendency is for a decrease to only 35 ppm
because other growth-factors are incorporated. Zinc supplementation
remains at about 150 ppm Zn.*

*Two methods were used for these investigations: 1) Direct
measurement of Cu enrichment of soils for 190 fields (64 breeders);
2) Pot and microplot experiments with $CuSO_4$ incorporation into the soil.
Soils analysed in 1979, 6 years, therefore, after the beginning of this
survey, indicated a definite enrichment for Cu (NH_4Ac-EDTA method) and an
even greater one for Zn. Averages of increases are respectively +1.23 ppm
and +2.96 ppm. In farms with the highest pig density, enrichments are:
+2.82 ppm for Cu and +9.06 ppm for Zn. Copper and zinc were also determined by 0.5 N,HNO3 and by nitroperchloric digestion. From our pot and
field experiments we believe it is dangerous to incorporate 500 kg/ha Cu
or Zn into Finisterian soils for cultures such as Italian rye-grass, maize,
winter-wheat, barley ($\geq$ 120 ppm Cu or Zn by EDTA method). If we take, for
example, a pig fattener with moderate pig density (70 pigs produced/ha/year
= 50 t slurry/ha/year) and with feeds containing Cu 125 ppm and Zn 150 ppm,
copper and zinc incorporated in the arable layer would reach 500 kg/ha
after two centuries; as we believe that the phytotoxicity of Cu and Zn are
complementary, the dangers could appear after only one century.*

INTRODUCTION

In Brittany pig breeding is particularly intensive.
This province has four 'Départements' with two million hectares
of Utilised Agricultural Surface (UAS) and a total of more than
4 300 000 pigs (sows + piglets + fattening pigs). Of these
four departments, Finistère was the subject of our particular
attention. Here are some characteristics:

UAS	460 000 ha
Pigs	3/ha
Cattle (Livestock units)	1.7/ha
A few communes have	5 to 10 pigs/ha
	(10 to 20 fattening pigs produced/ha/year)
Many fatteners have	35 pigs/ha
	(70 fattening pigs produced/ha/year)
A few breeder-fatteners	12 sows and progeny/ha
	(240 fattening pigs produced/ha/year)

Previous publications (Coppenet, 1974 and 1976; Duthion, 1976)
have pointed out that maximum rate of pig slurry per hectare is
about 50 t/year. This rate corresponds to 250 kg N/ha of which
175 kg is in the form of ammonium nitrogen, from 70 fattening
pigs produced/ha/year. In Brittany, the principal crops are,
forage maize, grain maize, cereals, Italian and English ryegrass, a few other cultivated plants but little permanent
grassland. Consequently pig slurry and cow slurry are spread
over ploughed soils.

Copper supplementation of feeds for fattening pigs was
often 125 ppm Cu five years ago but today the tendency is for
a decrease to only 35 ppm because other growth factors are
incorporated. Zinc supplementation remains at about 150 ppm Zn.

MATERIAL AND METHODS

1)   <u>Survey conducted with the collaboration of the 'Chambre
     d'Agriculture du Finistère'. (Professional organisation)</u>

This survey began in 1973 and concerned 64 breeders who

each submitted soil samples from 3 fields for analysis; accordingly, a total of 190 fields were analysed in 1973, 1976 and 1979. On these 64 farms (10 to 40 ha UAS), 54 have pigs (pigs only or pigs and cattle) with a high pig-density, and landspreading of slurries is intense: between 40 and 200 t/ha/ year. Crop production is followed up, but crops are not weighed. It was not possible to obtain data on copper sup- plementation for each farm.

2) Pot experiments and microplot experiments

Characteristics of Finisterian soils are: organic matter 5 to 6%, clay 12 to 18%, C.E.C. 12 - 15 m.éq 100 g, pH (H_2O) 5.8 to 6.2, depth of ploughing 25 cm. Pots contained 12 or 13 kg from the arable layer and we incorporated copper and zinc as sulphates separately or mixed on the following basis: 250 kg Cu or Zn/ha (100 mg Cu/kg of dry soil), 500, 1 000, 1 500 kg/ha. In order to neutralise subsequent acidification, we add calcium carbonate in sufficient quantity. A good rate of N, P and K is used. After 3 or 4 years crops become irregular and trials must be stopped.

First trial (granite soil) and second trial (schist soil) - 1973, Italian ryegrass - 1974, maize - 1975, winter wheat - 1976, spring barley.

Third trial (schist soil) - 1977, maize - 1978, winter wheat - 1979, spring barley.

Fourth trial (granite plots) - 1971, potatoes - 1972, maize - 1973, potatoes.

Fifth trial (granite plots) - 1976, winter wheat - 1977, maize - 1978, spring barley - 1979, potatoes.

RESULTS

1) Survey concerning 64 farms

Soils analysed in 1979, 6 years therefore after the beginning of this survey, indicated: very high enrichment for available P_2O_5 (citric and Olsen methods), moderate enrichment for exchangeable K_2O and MgO (Coppenet and Golven, 1979), a

definite enrichment for Cu and an even greater one for Zn,
(Table 1).

The higher the pig density/hectare, the higher the
build-up of P_2O_5, Cu, Zn. The method used for Cu and Zn is
extraction by normal ammonium acetate added to EDTA as disodic
salt 0.01 M and stirred for 2 hours (NH_4 Ac - EDTA Cu or Zn).
For 45 fields of this survey chosen at random, we determined,
other than EDTA Cu and Zn, Cu and Zn by 0.5 N,HNO_3 and by
nitroperchloric digestion, (Table 2).

It is necessary to continue these determinations for all
the fields but work is not yet finished. Exchangeable Cu by
normal ammonium acetate, which was also determined for some
samples or soils enriched by $CuSO_4$, is not a suitable method
as results are too uncertain. We believe that nitroperchloric
digestion extracts the totality of Cu and Zn added to the soils
by slurry. This view is corroborated by the following
calculations:

For Cu, + 2.18 ppm x 3 000 t/ha = 6 540 g Cu added over 6 years,
6.54 kg: 6 = 1.1 kg Cu/year, and if average rate of slurry
was 70 t/ha/year, we found 15 g Cu/t of slurry and 44 ppm
Cu in food.

For Zn, + 9.75 ppm x 3 000 t/ha = 29 250 g Zn added over 6
years, 29.25 kg: 6 = 4.87 Zn/year, and if average rate
of slurry was 70 t/ha/year, we found 69 g Zn/t of slurry
and 200 ppm Zn in food.

Ratios of EDTA and HNO_3 methods to nitroperchloric method

From Table 2 other conclusions can be drawn.

For Cu - 0.5 N, HNO_3 extracted 50% of nitroperchloric Cu in
1973 but 66% of added copper (1.44 / 2.18).
NH_4Ac-EDTA extracted 28% of nitroperchloric Cu in
1973 but 61% of added copper.
For Zn - 0.5 N, HNO_3 extracted 23% of nitroperchloric Zn in
1973 but 96% of added zinc.
NH_4Ac-EDTA extracted 9.3% of nitroperchloric Zn in
1973 but 40% of added zinc.

TABLE 1

RESULTS OF SOILS ANALYSES FOR Cu AND Zn (NH$_4$ Ac - EDTA METHOD)

	EDTA Cu ppm in dry soil			EDTA Zn ppm in dry soil		
	1973	1979	Increase	1973	1979	Increase
General average for the 190 fields	3.82	5.05	+1.23	3.19	6.15	+2.96
Maximum content	12.0	14.0	-	9.0	40.0	-
43 fields receiving less than 50 t/ha/year of slurry	3.13	4.12	+0.99	2.78	4.32	+1.54
88 fields receiving between 50 to 100 t/ha/year of slurry	4.26	5.26	+1.0	3.23	5.02	+1.79
29 fields receiving more than 100 t/ha/year of slurry	4.13	6.43	+2.3	4.09	13.36	+9.27
6 farms (18 fields) with pigs + bullocks 40 pigs produced/ha/year or 20 piggery places	4.68	5.97	+1.29	4.81	6.14	+1.33
17 farms (51 fields) with pigs + cow milch 60 pigs produced/ha/year or 30 piggery places	4.56	5.94	+1.38	2.96	5.74	+2.78
10 farms (30 fields) with pigs only 90 pigs produced/ha/year or 45 piggery places	4.43	5.71	+1.28	3.43	5.83	+2.40
4 farms (12 fields) with pigs only 260 pigs produced/ha/year or 130 piggery places	2.79	5.61	+2.82	4.75	13.81	+9.06

TABLE 2

RESULTS OF SOILS ANALYSES FOR THREE METHODS AND 45 FIELDS

| | Copper | | | | Zinc | | | |
| | Index in 1973 | ppm in dry soil | | | Index in 1973 | ppm in dry soil | | |
		1973	1979	Increase		1973	1979	Increase
Nitroperchloric digestion	100	13.22	15.40	+2.18	100	42.68	52.43	+9.75
0.5 N, HNO_3	50	6.59	8.03	+1.44	23	9.98	19.34	+9.36
NH_4 Ac - EDTA	28	3.71	5.04	+1.33	9.3	3.96	7.84	+3.88

Italian ryegrass cultivated on these soils and analysed at pasture stage has an average content of 8 ppm Cu/DM and of 30 ppm Zn/DM.

Maximum contents can reach 15 ppm Cu and 60 ppm Zn. Average copper content is not sufficient for cow-milch with high production but maximum contents can be dangerous for sheep.

2) Copper and zinc phytotoxicity

From the synthesis of the 5 experiments described above, we conclude that in Finistère:

- Phytotoxicity appears at approximately,

Cu alone	- 500 kg/ha	$\geq$ 120 ppm EDTA Cu
Cu + Zn	- 250 + 250 kg/ha	$\geq$ 70 ppm Cu + 50 ppm Zn (EDTA method)
Zn alone	- 500 kg/ha	$\geq$ 100 ppm EDTA Zn

- 50% yield reduction compared with check sample,

Cu alone	- 1 200 kg/ha	$\geq$ 250 ppm EDTA Cu
Cu + Zn	- 750 + 750 kg/ha	$\geq$ 200 ppm Cu + 200 ppm Zn (EDTA method)

Yellowing, together with the shrivelling of the lower parts of leaves, is a manifestation of copper toxicity in wheat; on maize, veins and plants turn purple or violet. Spring barley in 1979 from the fifth trial was manganese-deficient for pots with 500, 1 000 and 1 500 kg Zn/ha.

EDTA Cu / HNO_3Cu ratio for non-enriched soils is 0.6 to 0.7 (for Zn 0.4 to 0.5). This ratio increases with the addition of copper or zinc salts.

CONCLUSIONS

From our pot and field experiments we believe it is dangerous to incorporate 500 kg/ha Cu or Zn into Finisterian soils for cultures such as Italian ryegrass, maize, winter wheat, spring barley. Spreading slurry from fattening pigs is the reason for the large build-up of Cu and Zn in soils, but bullock slurry is also very rich in zinc.

If we take, for example, a pig fattener with moderate pig density (70 pigs produced/ha/year = 50 t slurry/ha/year) and with feeds containing Zn 150 ppm and Cu 35 ppm, zinc incorporated in the arable layer would reach 500 kg/ha after two centuries and copper would reach this same rate after eight centuries. But if Cu in feeds is 125 ppm the quantity of Cu added in soil will reach 500 kg/ha after only two centuries; as we believe that the phytotoxicity of Cu and Zn are complementary, the dangers could appear after only one century.

Obviously if pig density is higher with, for example, 140 pigs produced/ha/year (that is the case for some breeders) dangers can arise after 50 years.

Laboratory analyses of soils enable us to follow build-up for Cu and Zn and results obtained in Brittany from this survey confirm the theoretical predictions.

REFERENCES

Coppenet, M., 1974. L'épandage du lisier de porcherie. Ses conséquences agronomiques. Ann. Agron. 25, 403-423.

Coppenet, M., 1976. Utilizzazione agronomica delle deiezioni suine in 'Utilizzazione e valorizzazione dei liquami di porcilaia'. Rassegna suinicola internazionale. Camera di commercio, industria, artigianato e agricoltura. Reggio Emilia, 109-117 (A book of 199 pages) and in 'L'éleveur de porc'. 73, 81-85.

Coppenet, M. and Golven, J., 1979. Emploi rationnel des éléments fertili-sants contenus dans les déjections animales et en particulier dans les lisiers. Premiers résultats de l'enquête 'Lisier-Sol-Plante' en cours dans le Finistère. Colloque international in 'Cahiers du CENECA'. Paris 1979, N° 3218, 7 pages.

Duthion, C., 1976. Consequences of semi-liquid pig manure spreading. In 'Utilization of manure by land spreading'. EEC Seminar, Modena, Italy, (September 20-24, 1976), EUR 5672 e, 45-50.

A CONTRIBUTION TO THE ESTABLISHMENT OF SAFE COPPER LEVELS IN SOIL

Th.M. Lexmond

Department of Soils and Fertilisers,
Agricultural University, Wageningen, The Netherlands.

ABSTRACT

Soil pH and organic matter content are considered the main factors which modify the relation between the copper content of the soil and the incidence of toxicity to crop plants. When these factors are taken into account, the literature on copper toxicity shows a remarkable consistency. Toxicity is unlikely to occur on slightly to moderately acid soils when the amount of HNO_3 extractable copper does not exceed 20 to 30 mg/kg for each per cent of organic carbon present.

INTRODUCTION

The addition of copper salts to the rations for fattening pigs entails the production of manure with a high copper
content, which most often is spread on agricultural land.
Repeated application of such manure at rates recommended for
the major nutrients results in an accumulation of copper in
the top soil layers. Concern has arisen that the copper
content of soils may eventually reach levels which bring about
undesirable consequences. The Dublin Workshop on animal and
human health hazards associated with the utilisation of animal
effluents recommended therefore that 'Research should be undertaken to establish soil values in order to determine safe
levels of copper for plant growth and animal health' (Kelly,
1978).

This paper presents data which may facilitate the
interpretation of soil copper values, particularly with an eye
to the risk of toxicity to agricultural crops. To this effect
it contains evidence on how several soil factors affect the
relationship between the copper content of the soil and crop
performance, thus causing the seeming inconsistency of the
literature noted by Hartmans (1978) in his contribution to the
Dublin Workshop. For some of these factors sufficient information is available to warrant a quantitative approach, albeit
a rather simplified one.

In pursuit of the utmost clarity, attention is paid first
to water - plant and soil - water subsystems, before discussing
the results acquired from complete soil - water - plant
systems. This last section is followed by a critical evaluation of the literature data listed by Hartmans, supplemented
with some other published data.

COPPER TOXICITY IN WATER - PLANT SYSTEMS

Solution culture experiments have shown that the
occurrence of Cu phytotoxicity is determined by the activity

of the free cupric ions. The addition of complexing agents, which lower the Cu^{2+} ion activity, renders Cu less toxic (Majumder and Dunn, 1959; Dragun et al., 1976).

Consequently, the use of soil solution analysis to characterise the Cu status of the soil (as suggested by Bradford et al., 1971) is only correct when methods are employed which distinguish between free and complexed forms of Cu. Because an important part of the Cu dissolved in the soil solution or in a water extract of the soil is in a complexed form, the failure to make this distinction leads to erroneous conclusions: Majumder and Dunn observed toxicity to maize at a Cu concentration of 0.4 µmole/l, whereas Bradford et al. found the Cu concentration in saturation extracts of normal (i.e. non-polluted) California soils to have a median value of 0.5 µmole/l.

In the absence of complexing agents the toxicity of Cu in solution increases considerably with rising pH (Lexmond and Van der Vorm, to be published). Figure 1 presents some relevant results of an experiment in which maize seedlings were grown for 16 days in constantly flowing nutrient solutions containing different Cu concentrations, each at three pH levels. Toxicity of Cu affected most distinctly the morphology of the root system (root surface area) and the accumulation of P and Fe in the shoots. As the changes in root surface area are not easily quantified, the results for P-uptake are shown in Figure 1A. None of the Cu concentrations used reduced P-uptake at pH 4.0, 0.10 mg/l Cu had a significant effect at pH 5.8, and 0.20 but not 0.10 mg/l affected P-uptake at pH 4.7. An increase in pH was associated with a strong increase in the Cu content of the roots (Figure 1C), but had no such effect on shoot Cu (Figure 1B). Apparently, the occurrence of toxicity is related to the Cu content of the roots rather than the shoots (Figure 1D).

Other evidence on this matter is available from work by Hunter (1975). His results (Table 1) demonstrate that in-

creasing the pH increases Cu uptake by maize roots and
decreases subsequent growth.

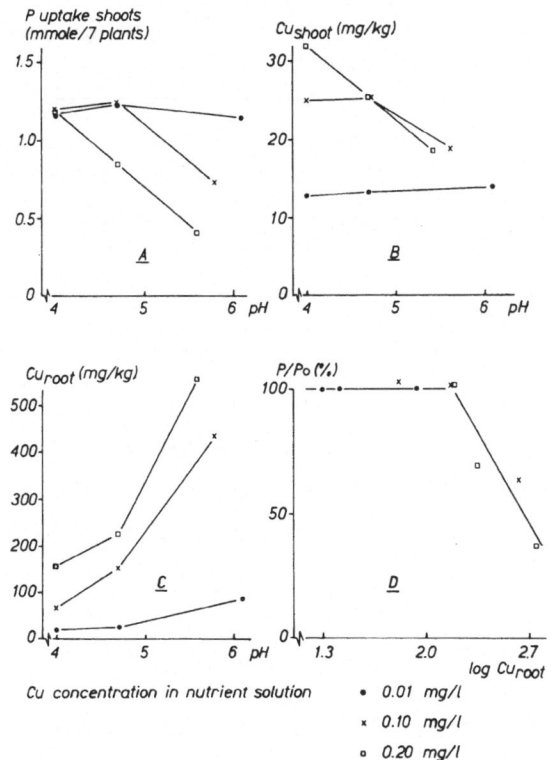

Fig. 1.   Modifying influence of solution pH on the effect of Cu on

A- phosphate accumulation by shoots of maize
B- the Cu content of maize shoots
C- the Cu content of maize roots

D- Relationship between the Cu content of maize roots and the
decrease in P accumulation by the shoots.

The same effect of pH has been observed in studies with
algae and fungal spores. Steemann Nielsen and Kamp-Nielsen
(1970) noted that at pH 5 about ten times as much Cu is
required than at pH 8 to obtain a comparably deleterious effect
to *Chlorella pyrenoidosa*. Lowering the pH from 8 to 5 strongly
reduced the amount of Cu taken up by the algal cells (Steemann

TABLE 1

COPPER UPTAKE AND SUBSEQUENT GROWTH OF MAIZE ROOTS TREATED FOR ONE HOUR
WITH NUTRIENT SOLUTIONS CONTAINING 8 mg/l Cu AT VARYING pH (FROM HUNTER, 1975)

Cu (mg/l)	pH	root Cu (μg/root)	24 hour root growth (cm ± se)
8	4.4	1.43	0.25 ± 0.03
8	5.0	2.50	0.12 ± 0.01
8	5.8	3.17	0.05 ± 0.02
0	5	0.23	3.33 ± 0.09

Nielsen et al., 1969). Biedermann and Müller (1952) found
that the inhibitory effect of Cu solutions on the germination
of *Alternaria tenuis* spores could be reduced or even overcome by
lowering the solution pH.

Apart from an increase in Cu uptake and toxicity with
increasing pH, several other interesting similarities exist
between the effects of Cu on algae, fungal spores and maize
roots. During an initial phase of several hours, in which
growth is already inhibited, the toxicity can be reversed by
rinsing with complexants which remove a portion of the Cu taken
up (Hunter, 1975; Somers, 1963). During this period Cu does
not penetrate into the symplasm (Steemann Nielsen et al., 1969),
so the site of its primary toxic effect must be located in the
apoplasm. Since this initial phase is also characterised by
a liberation of K^+ ions from the cells in amounts which cannot
be accounted for by regular ion exchange (Hunter, 1975; Wainwright and Woolhouse, 1977), the most plausible inference is
that Cu is bound to the plasmalemma, thereby altering its
retentive properties, inhibiting growth and possibly depressing
the rate of active nutrient uptake.

The uptake of Cu by fungal spores and plant roots follows the Freundlich adsorption equation (McCallan and Miller, 1958; Veltrup, 1976), thus reflecting the passive character of the binding process. The Cu content of maize roots in the experiment by Lexmond and Van der Vorm referred to before, can be described as a function of Cu concentration and pH by a Freundlich equation which has been modified to include a pH term:

$$\log Cu_{root} = 1.20 + 0.72 \log C_{Cu} + 0.36 \text{ pH} \qquad R^2 = 0.984 \qquad (1a)$$

or, substituting the measured Cu^{2+} ion activity values and rearranging:

$$\log Cu_{root} = 4.80 - 0.72 \text{ (pCu - 0.50 pH)} \qquad (1b)$$

In summary, the intensity of Cu toxicity to plants is related to the Cu content of the roots, which depends on both the Cu^{2+} ion activity and pH. These parameters must be known to characterise solutions with respect to the risk of Cu toxicity.

BEHAVIOUR OF COPPER IN SOIL - WATER SYSTEMS

In this section attention will be focused on the relation between the Cu content of the soil and the Cu^{2+} ion activity in the soil solution. In principle one may distinguish between two mechanisms by which the Cu^{2+} activity can be controlled in soils, namely precipitation and adsorption reactions. There is ample evidence to suggest that, under the conditions prevailing in the top layers of well aerated agricultural soils, the Cu^{2+} activity is governed by adsorption and not by precipitation of sparingly soluble compounds. In French vineyard soils, where Cu has increased to high levels during a period of 70 years, a high percentage of the Cu added remains isotopically exchangeable (Delas et al., 1960). Cu in these soils has been shown to be associated mainly with organic matter (Alloway et al., 1979). From the known solubility products of Cu minerals one would also infer that they are much too soluble to keep the Cu^{2+} activity low enough to prevent phytotoxicity in non-calcareous soils.

With adsorption being the activity controlling process, each addition of Cu to a soil will lead to an increase in Cu^{2+} activity, the rate of which depends, apart from the amount added, on the buffer power of the soil. Soils differ considerably in their buffer power towards Cu. As is clearly shown in Figure 2, the quantity of Cu that must be added to a soil to induce a certain increase in Cu concentration is soil dependent. The work of McLaren and Crawford (1973b) confirmed earlier observations by Wei (1959) that soil organic matter is much more effective than clay minerals in keeping the Cu^{2+} activity low. McLaren and Crawford (1973b) did recognise the high affinity of free manganese oxides for Cu, but because of its far greater abundancy organic matter is the dominant factor in soils (Figure 2). One might expect that different types of organic matter vary in their buffer power towards Cu. Indeed, Pratt et al. (1964) observed that subsurface samples of peat soils adsorbed less Cu/g organic carbon than surface soils. The organic matter of mineral soils showed the same adsorption characteristics as organic matter from surface peats. Therefore it seems justified for the present purpose to assume that the soil's buffer power is solely determined by its organic matter content and that the buffer power increases linearly with the organic carbon content.

Since the contribution of clay minerals to the soil's buffer power is relatively small, the use of the cation exchange capacity (CEC) as a predictor of the loading capacity of soils for Cu (EPA, 1978) may lead to Cu toxicity on clayey soils low in organic matter. The following data may serve to illustrate this point: 4.0 mg Cu per meq CEC caused 15% yield reduction in maize on a sandy soil with the CEC derived from organic matter, whereas 1.0 mg Cu per meq CEC brought about a decrease in yield of over 80% on a heavy clay soil under comparable conditions (Lexmond, unpublished results).

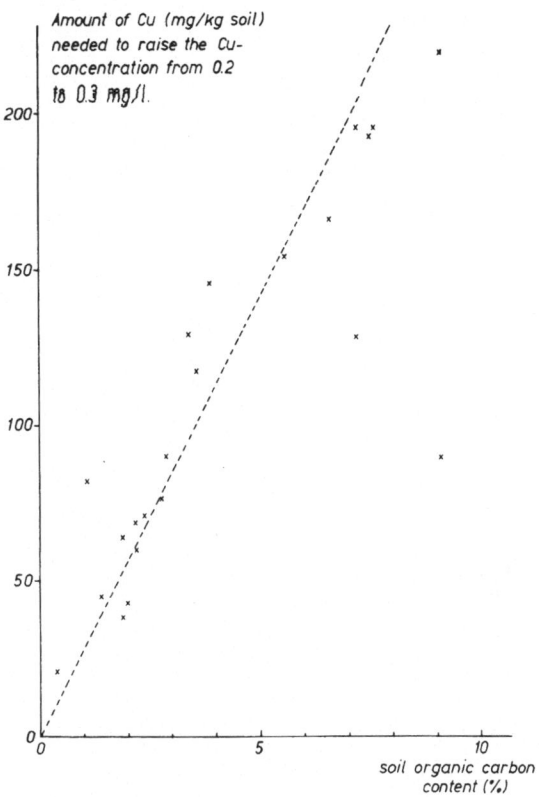

Fig. 2. Soil buffer power with respect to Cu as a function of the organic carbon content (derived from McLaren and Crawford, 1973a, b). The line has been forced through the origin.

Soil pH has a substantial effect on the buffer power of soils and soil constituents (McLaren and Crawford, 1973b), and as a result lowering the pH increases the Cu^{2+} activity. This effect is demonstrated by Figure 3, from which it appears that this effect occurs independently of the Cu content of the soil. Besides that, it also illustrates the adsorptive character of the Cu binding process, as the Cu^{2+} activity rises with each increment in the soil Cu content. The effects of both pH and Cu content can be described by a Freundlich equation with a pH term included:

$$pCu-CaCl_2 = 5.08 - 2.38 \log Cu-HNO_3 + 1.07 \ pH-CaCl_2 \quad R^2 = 0.989 \tag{2}$$

where $Cu-HNO_3$ represents the Cu content of the soil (mg/kg) as

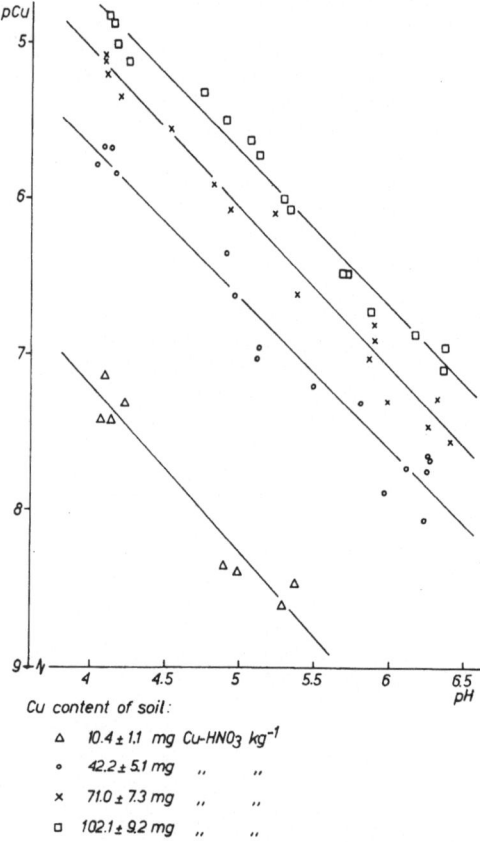

Fig. 3. Cu^{2+} activity in 10 mmole/l $CaCl_2$ as a function of pH at four soil Cu levels (from Lexmond, 1980).

measured by extraction with 0.43 mole/l HNO_3. This extractant was used because it removes the bulk of adsorbed Cu from the soil, without being so strong as to dissolve non-labile Cu fixed in lattice structures. It therefore seems well suited to yield a measure for adsorbed (or reversibly bound) Cu in soils. In addition, the efficiency of the HNO_3 extraction is independent of the soil pH, virtually constant over a wide range of soil Cu contents and not (or only slightly) affected by the organic matter content and CEC of the soil (Lexmond, 1980 and unpublished results). In view of the dominating influence of organic matter on the soil's buffer power, the amount of HNO_3 extractable Cu related to the organic carbon content (e.g. expressed as g Cu/kg C) might be of value when comparing the Cu status of different soils.

Summarising this section, it may be stated that the relationship between the quantity of adsorbed Cu in the soil and the Cu^{2+} ion activity in the soil solution depends on the buffer power of the soil, which is determined by the soil's composition and pH. Of the soil constituents, organic matter supplies the main contribution to the soil's buffer power.

COPPER TOXICITY IN SOIL - WATER - PLANT SYSTEMS

From the foregoing discussion it is apparent that no simple relationship can be expected to exist between the Cu content of a soil and the occurrence of toxicity to plants. The organic matter content of the soil and its pH have been recognised as major modifying factors.

Basically, the best way to deal with experiments in which Cu toxicity to soil grown plants is studied would be to measure the Cu^{2+} activity and pH in the soil solution and to relate the effects on plant growth to these parameters. By this procedure threshold values would be obtained for the Cu^{2+} activity in relation to pH for several crop species. Monitoring soils that receive Cu applications by the same methods would then indicate the development of unfavourable situations.

This approach was approximated by Lexmond (1980) in a field experiment pertaining to the effect of soil pH on Cu toxicity to forage maize. He estimated the Cu^{2+} activity and pH after equilibrating soil samples with 10 mmole/l $CaCl_2$ solutions, which served as an approximation for the soil solution. As with the results obtained in solution culture experiments by Lexmond and Van der Vorm, the relationship between the Cu^{2+} activity and crop response was found to depend on pH. The linear combination of pCu and pH (cf. equation 1b) which best predicted the effect on crop yield, and therefore could serve as a 'toxicity index', was calculated as (pCu - 0.55 pH). The regression equation describing relative yield (i.e. the ratio of yield (Y) to the yield of the O Cu treatment at the same pH level (Yo)) as a function of this 'toxicity index' reads:

$$Y/Yo = - 5.85 + 3.23 \ (pCu - 0.55 \ pH) - 0.38 \ (pCu - 0.55 \ pH)^2$$
$$R^2 = 0.955 \quad (3)$$

This relationship is presented in Figure 4. As Cu had a slightly stronger effect on P uptake than on yield, P uptake is also shown as a function of the 'toxicity index'. For reasons of comparison, the solution culture data discussed before are also included. The effect of Cu is somewhat more pronounced under field conditions than would appear from the solution culture experiment. An explanation might be found in the possibility that the formation of root hairs is more sensitive to high Cu^{2+} activities than the uptake mechanism for P. Root hairs play an important role in the uptake of immobile nutrients like P from soil, because of the extension of the root surface area they bring about. In solution cultures, however, any effect on root hair formation would escape observation because they are not formed at all under those conditions.

Having in view a comparison of results like those presented in Figure 4 with data reported in the literature, a different presentation is required. From equations 3 and 2 the soil Cu level can be calculated at which a 10% yield reduction can be expected:

$$\log Cu\text{-}HNO_3 = 0.58 + 0.22 \ pH\text{-}CaCl_2 \tag{4a}$$

As the soil Cu content is expressed more meaningfully on the basis of organic carbon, equation 4a is converted into:

$$\log Cu/C = - 0.32 + 0.22 \ pH\text{-}CaCl_2 \tag{4b}$$

with Cu/C being used as the abbreviation for g $Cu\text{-}HNO_3$/kg organic carbon.

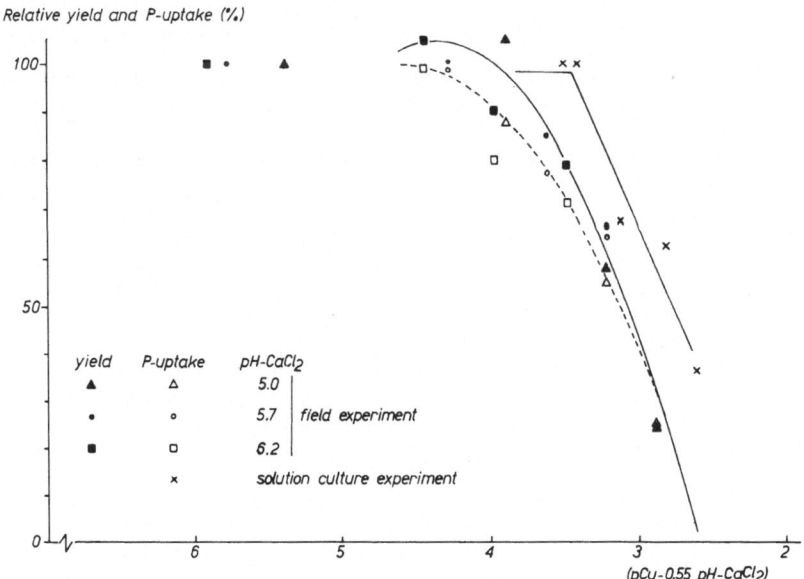

Fig. 4. Relative yield and P uptake as a function of the 'toxicity index' pCu - 0.55 pH as observed under field conditions (Lexmond, 1980) and in solution culture (Lexmond and Van der Vorm, to be published).

Calculated in the same way, the soil Cu level that brings about a 10% reduction in P uptake is given by:

$$\log Cu/C = -0.43 + 0.22 \; pH\text{-}CaCl_2 \tag{5}$$

Lexmond and De Haan (1980) presented results of a pot experiment with maize grown on 7 soils, each at 5 Cu levels. From their data it appears that a 10% decrease in P uptake occurs when:

$$\log Cu/C = -0.37 + 0.21 \; pH\text{-}CaCl_2 \tag{6}$$

which is in fair agreement with equation 5 obtained from the field experiment.

Equations 4b, 5 and 6 are shown in Figure 5, together with the results of a pot experiment with oats on 6 different soils, each at 6 Cu levels, carried out at the Institute for

Soil Fertility, Haren, the Netherlands (courtesy Ir. S. de
Haan). From this comparison it is evident that oats are less
susceptible to Cu toxicity than maize. The relative tolerance
of oats to Cu has also been observed by Forster (1953), Kühn
and Schaumlöffel (1961), Purves (1969) and Roth et al. (1971).

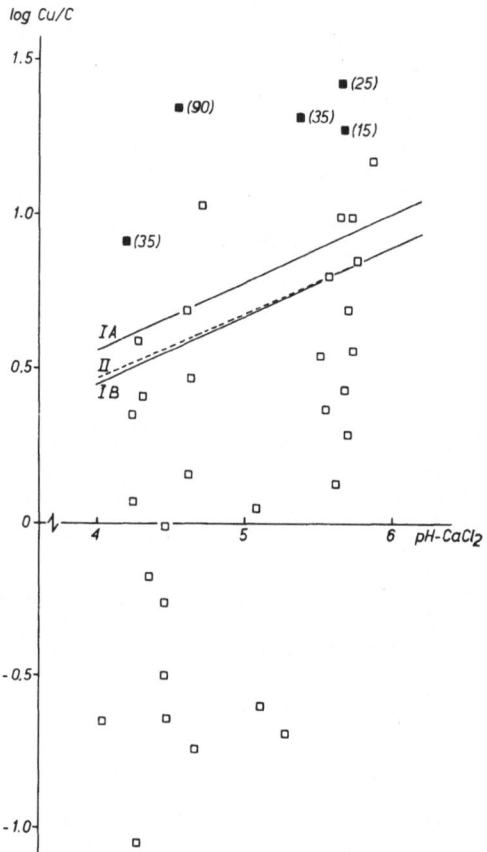

Fig. 5. Critical levels of Cu/C (g/kg) in relation to soil pH.

    I   field experiment with forage maize, Lexmond (1980)
   a 10% yield reduction
   b 10% reduction in P uptake
  II   pot experiment with maize, Lexmond and De Haan (1980),
      10% reduction in P uptake.

Squares refer to results of a pot experiment with oats; open
symbols denote no toxic effect, closed symbols denote toxicity.
Values in parentheses indicate % yield reduction. Courtesy Ir.
S. de Haan, Institute for Soil Fertility, Haren, the Netherlands.

PUBLISHED DATA CONCERNING Cu TOXICITY IN SOIL - WATER - PLANT SYSTEMS

Hartmans (1978) listed a number of data on critical Cu levels in soil in relation to crop growth. The purpose of this section is to evaluate these results and to compare them with the results presented in the previous section. It should be stressed that this part can by no means be considered exhaustive.

1) Reuther and Smith (1954)

With the soil pH near 5, citrus seedlings begin to show Cu toxicity when the total Cu level reaches about 1.6 mg of Cu per meq of cation exchange capacity (CEC). At twice this level of total Cu in relation to CEC mild to severe toxicity is likely to occur when the soil pH is in the range between 6 and 4. Growth of seedlings is progressively depressed as the soil becomes more acid.

Remarks: 1. pH in this paper is $pH-H_2O$, $pH-CaCl_2$ will be 0.2 to 0.5 units lower.

2. the CEC of Florida citrus soils is largely derived from the organic matter they contain. The CEC, as measured with neutral ammonium acetate, amounts to 1.92 meq/g organic matter (Ensminger, 1944). Assuming that the organic matter contains 58% organic carbon, 1 meq of CEC is equivalent to 0.30 g C.

3. the sandy Florida soils are very low in Cu by nature, so the quantity of Cu which is fixed in lattice structures can be neglected. It is assumed that 90% of the Cu added is extractable with HNO_3.

2) Spencer (1966)

Growth of citrus seedlings is reduced after application of 100 mg Cu per kg soil. The soil contains 1 - 1.5% organic

matter and has a CEC of circa 2.5 meq per 100 g. Increasing
rates of phosphate fertiliser decrease the toxic effect of Cu,
partly because of the accompanying increase in pH-H_2O (from
5.5 to 6.2), but also because the Cu induced inhibition of P
uptake is counteracted. The yield reduction averages 40%.

Remarks: 1.   the soil used is a Florida sandy soil and con-
                sequently remarks 1.2 and 1.3 apply to it. A
                value of 2.5 meq/100 g for the CEC indicates an
                organic matter content of 1.3%, which is well
                within the range stated.

3)   Walsh et al. (1972)

This paper is often referred to as an illustration of
what is called the 'special susceptibility of legumes to Cu
poisoning'. It therefore seems appropriate to give the original
data (Table 2).

TABLE 2

EFFECT OF FORM AND RATE OF APPLIED Cu ON THE YIELD OF BEANS (*Phaseolus
vulgaris*), (WALSH ET AL., 1972).

Carrier	Rate (kg Cu/ha)	Pod yield (kg/ha)
–	–	8 176
$CuSO_4$	18	9 374
	54	8 299
	162	8 042
	486	1 938
$Cu(OH)_2$	15	10 315
	45	7 896
	135	7 011
	405	5 376
	LSD (0.10)	1 355

Significant yield reductions occur at the highest
application rates only, soil additions up to 160 kg Cu/ha

apparently have no effect on crop growth. The soil contains 0.7% organic matter and has a pH-H$_2$O of 6.7.

Remarks: 1.   it is assumed that the Cu applied was mixed into 2.5 to 3.2.10^6 kg of soil and that 90% is extractable with HNO$_3$.

4)   Purves and MacKenzie (1973)

The effect reported in this paper concerns inhibited growth of beans (*Phaseolus vulgaris*) following application of municipal compost to soil. From soil and plant analysis data the authors conclude that B toxicity is likely to be the causative factor. In the discussion section of the paper reference is made to results of a pot experiment originally published by Purves (1965), which would indicate that levels above 30 mg EDTA-extractable Cu/kg soil bring about toxicity in clover. The interpretation of this value is rendered impossible, because no information is given on soil composition and pH.

5)   Henkens (1975 and unpublished results)

Two series of pot experiments have been carried out in 1974 and 1975 by students of the Agricultural University, Wageningen, the Netherlands, to obtain information on acceptable Cu levels in soil on behalf of the extension service. The results of the 1974 experiments have been published by Henkens (1975). Yield of maize, beans (*Phaseolus vulgaris*) and sugar beets is reduced by approximately 10% when the Cu-HNO$_3$ level reaches 67 mg/kg. Spinach yield is already depressed by 10% at a Cu-HNO$_3$ level of 55 mg/kg. The soil is a sandy soil with 2.5% loss on ignition, which corresponds to 1.0% organic carbon. The pH is 4.8 as measured in KCl (ca. 5.3 in CaCl$_2$). The 1975 experiments have not been published, but their results are the unpublished data referred to by Hartmans (1978). Addition of 125 mg Cu/kg soil does not cause toxicity in maize and sugar beets on soil C (4.1% org. C, pH-KCl 5.5). A slight toxicity occurs to maize, but not to sugar beets, on soil B

(2.2% org. C., pH-KCl 4.7) at an estimated Cu-HNO$_3$ level of 100 mg/kg. Addition of about 50 mg Cu/kg soil induces toxicity to both crops on soil A, which is the same soil as used in 1974.

Remarks: 1. soils A, B and C have also been used for the 1978 pot experiment reported by Lexmond and De Haan (1980), the results of which are represented by equation 6.

6)   Kühn and Schaumlöffel (1961)

A slight but significant yield reduction in spring wheat is observed following the application of 750 mg Cu as CuSO$_4$ to 6 dm^3 of a Cu deficient Oldenburg soil. The soil pH is 4.9 - 5.0 as measured in KCl. In another experiment about 2 000 mg Cu added to 6 dm^3 of another Oldenburg soil caused a negative effect on the growth of oats at a pH-KCl of 5.0.

Remarks: 1. from the reported data on loss on ignition and the mineral fraction < 20 µm, the organic carbon content of the soils used is estimated at 1.2 and 2.2%, respectively.

2. according to Scharrer and Schaumlöffel (1960), 6 dm^3 of soil corresponds to 6.5 and 6.0 kg, respectively.

3. pH-CaCl$_2$ is assumed to be 0.4 to 0.7 units higher than pH-KCl.

7)   Roth et al. (1971)

Addition of 128 µmoles of Cu/g of peaty muck soil (33% organic carbon, final pH-H$_2$O at saturation 6.1 to 6.4) causes a slight but not significant decrease in dry matter production in oats. 64 µmole/g induces a significant depression of the P content of soyabean tops. Both yield and P content of soya-bean are reduced by 128 µmole/g.

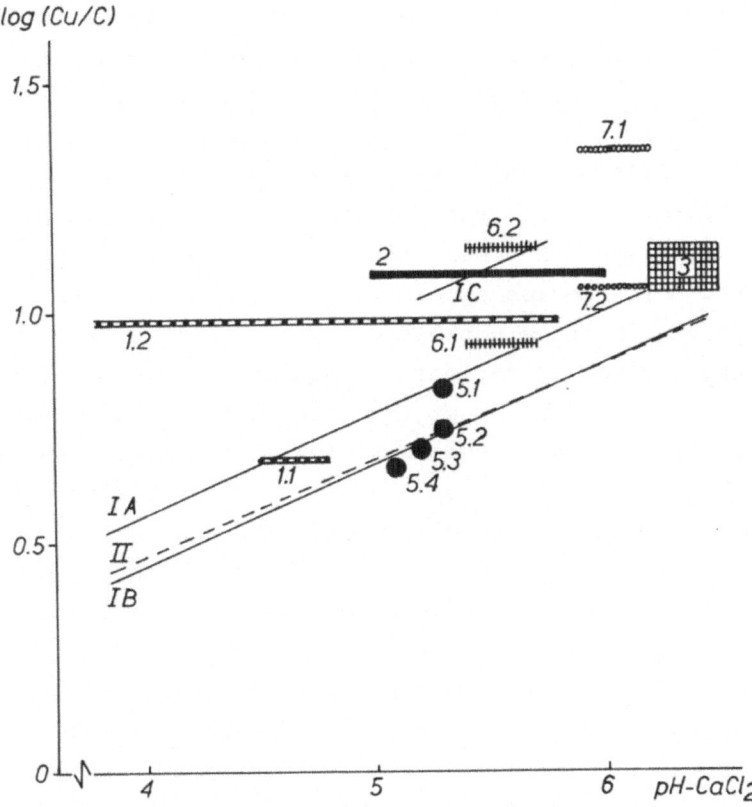

Fig. 6. Critical levels of Cu/C (g/kg) in relation to soil pH.

Reuther and Smith (1954), pot experiment with citrus seedlings
1.1 beginning toxicity
1.2 mild to severe toxicity, depending on pH
Spencer (1966), pot experiment with citrus seedlings
2. 40% yield reduction
Walsh et al. (1972), field experiment with beans
3. beginning toxicity
Henkens (1975 and unpublished results), pot experiments
5.1 approximately 10% yield reduction in maize, sugar beet and
     spring wheat
5.2 10% yield reduction in spinach
5.3 yield reduction in maize and sugar beet (soil A)
5.4 yield reduction in maize (soil B)
Kühn and Schaumlöffel (1961), pot experiments
6.1 significant yield reduction in spring wheat
6.2 yield reduction in oats
Roth et al. (1971), pot experiment
7.1 significant decrease in P uptake by soyabean
7.2 slight, but not significant effect on oat yield

Ia, Ib and II as in Fig. 5;
Ic 40% yield reduction in maize grown under field conditions
(Lexmond, 1980)

The data mentioned in this section are displayed graphically in Figure 6, and compared with the data obtained with maize, which were discussed in the previous section. The general pattern shows a remarkable consistency, thus indicating that the simplified model derived in this paper accounts for the most important modifying factors. The model did not include differences between plant species in their sensitivity to Cu, but most species appear to be comparable in this respect, oats, however, being a notable exception to this rule (Figure 5). There are no indications for a special susceptibility of legumes. Another factor not accounted for is the alleviating effect of high rates of phosphate fertiliser, noted by Reuther and Smith (1954) and Spencer (1964b). From solution culture experiments (Lexmond, unpublished) it appears that phosphate does not prevent damage to the root system of maize, but high P levels in soil may, to some extent, enable the plant to overcome the consequences of a reduced uptake capacity for P. It should also be mentioned that Reuther and Smith (1954) observed that Cu accumulated over a period of decades was less toxic than Cu freshly added to soils. All data presented here pertain to situations where Cu was added as a soluble salt. The two field experiments (Walsh et al., 1972; Lexmond, 1980) give no indication of a reduced toxicity, but the possibility of a fixation process active on a longer term cannot be precluded. One might for example consider a more important role for the (hydrous) oxides of Mn, Fe and Al. In case the HNO_3 extraction does not completely reflect such a decrease in Cu^{2+} activity with time, the data summarised in Figure 6 would overestimate the risk of Cu toxicity. On the other hand, it seems sensible to include a safety margin between the critical Cu levels indicated and the levels which can be considered acceptable. Some plant species (of current or potential interest to agriculture), not included in the studies referred to, may be more sensitive to Cu. Much remains to be learned about effects of combinations of heavy metals. Although the actual establishment of such a safety margin is a political matter, the results discussed indicate that in slightly to moderately acid soils

(pH-CaCl$_2$ 6.5 to 4.5) Cu phytotoxicity is unlikely to occur
when Cu-HNO$_3$ does not exceed 20 to 30 mg/kg for each per cent
of organic carbon present.

REFERENCES

Alloway, B.J., Gregson, M., Gregson, S.K., Tanner, R. and Tills, A., 1979.
Heavy metals in soils contaminated from several sources including
sewage sludge. Proc. Int. Conf. Management and Control of Heavy
Metals in the Environment, London, pp 545-548.

Biedermann, W. and Müller, E., 1952. Die Inaktivierung des gelösten
Kupfers (II) in Fungiziden. Phytopathol. Z. 18, 307-338.

Bradford, G.R., Bair, F.L. and Hunsaker, V., 1971. Trace and major element
contents of soil saturation extracts. Soil Sci. 112, 225-230.

Delas, J., Delmas, J. and Demias, C., 1960. Etude par dilution isotopique
du cuivre incorporé dans les sols depuis 70 ans. C.R. Acad. Sci.
250, 3867-3869.

Dragun, J., Baker, D.E. and Risius, M.L., 1976. Growth and element
accumulation by two single-cross corn hybrids as affected by copper
in solution. Agron. J. 68, 466-470.

Ensminger, L.E., 1944. Base exchange capacity of some Florida soils.
Soil Sci. Soc. Fla. Proc. 6, 166-168.

EPA, 1978. Sludge treatment and disposal. EPA-625/4-78-012. Environmental Research Information Center, Cincinatti, Ohio, USA.

Forster, W.A., 1953. Some effects of metals in excess on crop plants
grown in soil culture. V. Effects of copper on mineral status of oat,
rye, barley, sugar beet and kale. Ann. Rep. Long Ashton Res. Sta.
1952, 204-213.

Hartmans, J., 1978. Identifying the priority contaminants: toxicological
aspects of animal effluents. EUR 6009- Animal and human health
hazards associated with the utilization of animal effluents. Ed.
W.R. Kelly, pp 35-56.

Henkens, Ch. H., 1975. Zuiveringsslib in de landbouw. Bedrijfsontwikke-
ling 6, 98-103.

Hunter, R.B., 1975. Copper toxicity to roots of corn (*Zea mays*). Ph. D.
thesis, Utah State University, Logan, Utah, USA, pp. 102

Kelly, W.R. (Ed.), 1978. Animal and human health hazards associated with
the utilization of animal effluents, EUR 6009.

Kühn, H. and Schaumlöffel, E., 1961. Über die Wirkung hoher Kupfergaben
auf das Wachstum von Getreide. Landwirtsch. Forsch. 14, 82-90.

Lexmond, Th.M., 1980. The effect of soil pH on copper toxicity to forage
maize grown under field conditions. Neth. J. agric. Sci. 28, 164-183

Lexmond, Th.M. and De Haan, F.A.M., 1980. Toxicity of copper. EUR 6633 -
Effluents from livestock, Ed. J.K.R. Gasser, pp 410-418.

Majumder, S.K. and Dunn, S., 1959. Modifying effect of EDTA on the copper
toxicity to corn in nutrient solution. Plant Soil 10, 296-298.

McCallan, S.E.A. and Miller, L.P., 1958. Innate toxicity of fungicides.
Adv. Pest. Control Res. 2, 107-134.

McLaren, R.G. and Crawford, D.V., 1973a. Studies on soil copper I. The
fractionation of copper in soils. J. Soil Sci. 24, 172-181.

McLaren, R.G. and Crawford, D.V., 1973b. Studies on soil copper II. The
specific adsorption of copper by soils. J. Soil Sci. 24, 443-452.

Pratt, P.F., Bair, F.L. and McLean, G.W., 1964. Nickel and copper
chelation capacities of soil organic matter. Trans. 8th Inter. Congr.
Soil Science, Bucharest, Romania, 1964, Vol 3, pp 243-248.

Purves, D., 1965. The effect of high levels of EDTA-extractable copper in
soil on growth of clover. Exp. Work Edinburgh School of Agriculture
1964: 3.

Purves, D., 1969. Effects of high levels of EDTA-extractable copper on the
uptake of copper by oats and clover. Exp. Work Edinburgh School of
Agriculture 1968: 2-3.

Purves, D. and MacKenzie, E.J., 1973. Effects of application of municipal
compost on uptake of copper, zinc and boron by garden vegetables.
Plant Soil 39, 361-371.

Reuther, W. and Smith, P.F., 1954. Toxic effects of accumulated copper in
Florida soils. Soil Sci. Soc. Fla. Proc. 14, 17-23.

Roth, J.A., Wallihan, E.F. and Sharpless, R.G., 1971. Uptake by oats
and soybeans of copper and nickel added to a peat soil. Soil Sci.
112, 338-342.

Scharrer, K. and Schaumlöffel, E., 1960. Uber die Kupferaufnahme durch
Sommergetreide auf Kufpermangelböden. Z. Pflanzenernähr., Düng.,
Bodenkde 89, 1-17.

Somers, E., 1963. The uptake of copper by fungal cells. Ann. Appl. Biol.
51, 425-437.

Spencer, W.F., 1966. Effect of copper on yield and uptake of phosphorus
and iron by citrus seedlings grown at various phosphorus levels.
Soil Sci. 102, 296-299.

Steemann Nielsen, E., Kamp-Nielsen, L. and Wium-Andersen, S., 1969. The
    effect of deleterious concentrations of copper on the photosynthesis
    of *Chlorella pyrenoidosa*. Physiol. Plant. 22, 1121-1133.

Steemann Nielsen, E. and Kamp-Nielsen, L., 1970. Influence of deleterious
    concentrations of copper on the growth of *Chlorella pyrenoidosa*.
    Physiol. Plant. 23, 828-840.

Veltrup, W., 1976. Concentration dependent uptake of copper by barley
    roots. Physiol. Plant. 36, 217-220.

Wainwright, S.J. and Woolhouse, H.W., 1977. Some physiological aspects of
    copper and zinc tolerance in *Agrostis tenuis* Sibth.: cell elongation
    and membrane damage. J. exp. Bot. 28, 1029-1036.

Walsh, L.M., Erhardt, W.H. and Seibel, H.D., 1972. Copper toxicity in
    snapbeans (*Phaseolus vulgaris* L.). J. Environ. Quality 1, 197-200.

Wei, L.S., 1959. The chemistry of soil copper. Diss. Abstr. 19, 2712.

COPPER AND ITS UTILISATION IN DANISH AGRICULTURE*

A. Dam Kofoed

State Experimental Station, Vejen, Denmark.

ABSTRACT

Use of copper in modern agriculture needs more and more attention. Copper is used as an additive for feeding stuffs and for fertilisers. Crop uptake of copper is small and when farmers use very copper-enriched slurry, manure or sludge there will be a great positive copper balance in the soil. Under such conditions the copper content of the crops will increase and this can cause toxicity in animals, especially in sheep. The paper reviews the situation in Danish agriculture in particular.

*Previously published in Fertilizer Research, 1, 63-71, 1980, Martinus Nijhoff.

Increasing attention has been paid in recent years to the role of copper in agriculture. Copper is an essential nutrient for plants and animals. A lack of copper may result in physiological complications and excessive amounts of copper may cause toxicity. Copper is increasingly being added to feeding stuffs as a stimulant for pigs and added to fertilisers in order to avoid deficiency and so the amount of copper circulating in some agricultural systems has increased in recent years. A balance sheet of copper supplied and removed in Danish agriculture is presented in Table 1. This shows that about 300 g/ha more copper is applied than is needed to cover crop needs and leaching losses. While this amount would cause no risk of toxicity there are areas where large amounts of Cu-enriched manure is used and also, sometimes, Cu-enriched fertiliser, and, in the long run, some care with regard to copper additions to the soil needs to be exercised.

TABLE 1

SUPPLY AND REMOVAL OF COPPER (Cu, g/ha/yr) IN AGRICULTURAL SOILS IN DENMARK[a]. AVERAGE 1970 - 75

Item	Supply	Item	Removal
Commercial fertiliser	215	Crops	50
Animal manure	150	Leaching	5
Precipitation	18	Surplus in soil	330
Agricultural lime	2		
Total	385		385

[a] Personal communication K Skriver, Landskontoret for Planteavl, Viby, Jutland

COPPER IN SOIL

Copper is found:

1) in the crystal lattices of minerals,
2) adsorbed on clay and humus,
3) as chelates,

4) in ionic form in the soil solution.

Much of the copper in humus complexes and adsorbed on soil
colloids is not in equilibrium with the soil solution and must
be characterised as 'fixed' or not readily available.

Soil Cu content varies much with soil type. Expressed
as available Cu in air dried soil (Henriksen, 1957) the
average value in Jutland was 1.8 mg/kg in 1960 - 64 but by
1975 - 77 it had risen to 3.0, varying from 2.6 in the east to
3.1 in the north and 3.2 in the west (Andersen, 1978). Results
of the latest survey (Frederiksen, 1971) for variation of
classification 1970 - 71, comprising 2 634 analyses, are given
in Table 2. Seventy-five percent of the soil samples had Cu-
contents ranging between 1.1 and 5 mg/kg.

TABLE 2
CLASSIFICATION OF 2 634 SOIL ANALYSES FOR COPPER 1970 - 71

Available Cu, mg/kg	0.0-1.0	1.1-2.5	2.6-5.0	5.1-10.0	11-25	26-50	51-100
Number	171	952	1 026	408	66	9	2
Percentage	6.5	36.1	39.0	15.5	2.5	0.3	0.1

In Denmark, available Cu is determined by the complexone
method in which the soil is extracted with 0.02 M EDTA
(Henriksen, 1957).

Danish field experiments have indicated that, normally,
no yield response to applied Cu would be expected at soil Cu
contents above 3 mg/kg for black sandy soil or above 1 - 2
mg/kg for ordinary arable land.

Theoretically, application of 10 kg $CuSO_4$/ha would raise
soil Cu content by one unit but in practice the increase is no
more than 0.5 - 0.9.

Various factors, e.g. soil type, pH and crop, affect the optimum Cu content. The formerly used dosage of 20 - 50 kg/ha copper sulphate (5 - 12.5 kg/ha Cu) has been sufficient to eliminate copper deficiency in barley and to have an apprec- iable residual effect.

In using Cu-enriched fertiliser, the farmer supplies considerable amounts of copper and this results in a significant net addition of copper on a national basis (Table 1). Accept- able soil Cu levels in the form of $Cu-HNO_3$ or Cu-EDTA (Hartmans, 1978) are: legumes 30, arable crops 50 and grass 80 mg Cu $(HNO_3$ or EDTA)/kg.

It is said that soil Cu levels above 50 mg/kg decrease the population of earthworms and limit their activity (Hartmans, 1978).

COPPER AND PLANT PRODUCTION

Crops take up quite small amounts of copper, amounting to only about 20 - 100 g Cu/ha. The concentration in plants varies from 2 to 20 mg/kg on a dry matter basis, and is most commonly around 5 - 10 mg/kg. Depending on plant species, concentrations above 20 mg/kg may indicate toxicity.

Copper toxicity in plants is said to be caused by the ability of copper to replace other ions from physiologically important sites within the plant. A special symptom of excess- ive Cu is reduced root growth. As with iron deficiency, the first sign of Cu toxicity is chlorosis.

Some authors say that legumes are especially sensitive to Cu poisoning and this is shown in reduced yield (Hartmans, 1978). At somewhat higher Cu contents the yield of crops such as beet, maize and wheat are also decreased. Yield reduction may be expected at copper contents in dry matter of around 15 mg/kg which is not much above normal.

Copper toxicity will vary with soil type, pH and total Cu and is rather rare in agriculture but it might be encountered in the surroundings of copper mines or in soils which have received frequent applications of copper salts, as has been reported from France where Bordeaux mixture has been used in vineyards for many years. Cu toxicity may be more likely to occur on acid soils in which Cu fixation is less than in neutral soils and, thus, is more available. The application of copper-contaminated sludge and, as mentioned above, the use of animal manures containing Cu might result in toxic levels in the soil.

One of the considerations in assessing Cu availability, deficiency and toxicity is the interaction of Cu with other minerals, e.g. Mo, SO_4, Zn, Cd, Fe.

COPPER TOXICITY IN ANIMALS

Domestic animals vary much in sensitivity to copper poisoning. In the Netherlands 225 - 250 mg/Cu/kg was accepted for pig feeding. On the other hand sheep are very sensitive, and intake over a long period of fodder containing 15 - 20 mg/kg of Cu might result in accumulation of copper in the liver with critical consequences (Hartmans, 1978).

It must be borne in mind in this connection that plant-available copper in the soil may also be taken up by animals. Several cases are reported from the Netherlands where sheep were poisoned after grazing in areas where copper enriched pig slurry had been spread. Grazing cows and sheep, especially in very dry or very wet periods, are liable to ingest along with the grass an amount of soil equal to 10% or more of the daily intake of dry matter (Hartmans, 1978). The same author mentions a level of 15 mg/Cu/kg dry matter as being acceptable for animals and for human beings.

From Australia, cases of copper poisoning in sheep on natural grazings have been reported. Some plant samples from areas containing much copper in the soil showed 50 - 60 mg/Cu/ kg dry matter.

Elsewhere, where soil Cu contents are normal, Cu toxicity in sheep may be caused by low molybdenum levels. Copper poisoning depends not only on Cu content of the fodder but also on the content of Zn, Fe, Ca and Mo, mineral sulphate and, possibly, other compounds (Batey et al., 1972).

Other ruminants, i.e. cattle, are less sensitive to copper poisoning. Though the views of scientists differ, it is generally accepted that a level of 50 mg/Cu/kg dry matter for continuous intake is tolerable. This level seems rather high. The order of Cu tolerance in domestic animals from least to most tolerant is: sheep, cattle, pigs, poultry.

COPPER IN FODDER

In 1955, Barber et al. discussed the effect of copper sulphate on uptake and utilisation of pig feed. A supplement of 250 mg/Cu/kg dry matter had a good effect, especially at the beginning of growth. Since then, experiments with copper carbonate, copper oxide and copper chloride have shown similar effects.

Nowdays it is usual in the Netherlands to supplement pig feed with up to 225 mg/Cu/kg, which, added to the original Cu-content in the fodder makes a total of 250 mg/kg (De Haan et al., 1976).

Danish investigations (Hansen et al., 1974) conclude that 125 mg/Cu/kg, equivalent to 100 - 250 g copper sulphate over the growing period per 20 - 90 kg pig, should be sufficient. This supplement resulted in a saving of 11 kg feed per pig and the growing period was reduced by 5 days. Moreover Nielsen et al., 1979), the pigs' Cu-need is covered by 6 mg/kg. An example of a Danish feed for fattening pigs is 'DLG-svine foder-17' which contains 117 mg Cu/kg dry matter and is suitable for use with other fodder containing 25 mg/Cu/kg.

As Cu utilisation by animals is rather low, at around 2 - 6% of the intake, the Cu supplement must result in the manure being rich in Cu.

CONTENT OF COPPER IN MANURE

Different types of animal manures and sludges were analysed at Askov Experimental Station (Kjellerup and Søndergaard Klausen, 1975; Kjellerup and Lindhard, 1977).

Results of the Cu-determination are shown in Table 3.

TABLE 3

ANALYSES OF ANIMAL MANURE AND SEWAGE SLUDGE

Item	Number of samples	DM %	Sand %	Cu, mg/kg	
				In manure	In DM
Cattle manure	50	26.7		9	34
Pig manure	17	24.4		21	86
Poultry manure	28	43.5		30	69
Slurry, cattle and pig	33	8.3		4	48
Slurry, pig	21	6.8		18	265
Slurry, cattle	117	9.2		4	43
Sludge, household	3	41.0	27.5	46	113
Sludge, industry	3	23.9	7.4	353	1 477

Three samples of cattle slurry, not included in the figures of Table 3, contained 35 - 60 mg/Cu/kg on a wet basis (380 - 650 mg/kg dry matter). In all those cases animals passing to and from the cowshed had to wade through liquid containing a solution of copper sulphate.

The Cu-content of some Danish Cu-enriched commercial fertilisers is shown in Table 4.

TABLE 4

CONTENT OF COPPER IN CU-ENRICHED FERTILISER

Formula	Cu, %
PK 0 - 5 - 12 + Mg + Cu	0.2
PK 0 - 5 - 13 + Cu	0.4
PK 0 - 4 - 21 + Mg + Cu	0.2
PK 0 - 7 - 18 + Mg + Cu	0.2
NP 26 - 4 - 0 + Mg + Cu	0.1
NPK 14 - 4 - 17 + Mg + Cu	0.1
NPK 18 - 5 - 12 + Mg + Cu	0.2
NPK 23 - 3 - 7 + Mg Δ÷	0.1

Twenty-one percent of all the fertiliser used in Denmark contain Cu, the most popular ratios being 0 - 5 - 12 and 23 - 3 - 7.

COPPER APPLIED TO THE FIELD WITH ANIMAL MANURE

Animal manure has for centuries played an important part in the nearly closed copper cycle of most farms. Copper supplied in this way has been sufficient to meet crop requirements except on the black sandy humus soils which occur especially in Jutland and where crop failure was a frequent occurrence until 1931 when it was shown to be caused by Cu deficiency. Long before that there had been a feeling that copper was needed in soil for crop production. It is therefore of interest to know how much copper may be applied to the soil in different amounts of manure and this is shown in Table 5, for slurry and in Table 6 for solid manure.

Slurry having a Cu concentration of 50 mg/kg in dry matter and applied at 50 t/ha adds 225 g Cu with cattle slurry and 175 g/ha with pig slurry. Since the copper in slurry is almost as available as that in copper sulphate, 50 t/ha of slurry easily covers the crops' need for copper. With 50 t/ha solid manure, considerably more copper is applied.

TABLE 5

COPPER ADDITIONS (Cu, g/ha) TO SOIL FROM SLURRY

Cu in slurry DM, mg/kg	Cattle slurry, t/ha with 9% DM			Pig slurry, t/ha with 7% DM		
	50	100	200	50	100	200
50	225	450	900	175	350	700
100	450	900	1 800	350	700	1 400
250	1 125	2 250	4 500	875	1 750	3 500
500	2 250	4 500	9 000	1 750	3 500	7 000
1 000	4 500	9 000	18 000	3 500	7 000	14 000

If copper is applied in excess of crop needs, soil Cu content increases because, as with other heavy metals, only small amounts, in the region of 8 - 19 g/ha/y are leached (Kjellerup and Dam Kofoed, 1979). Analysis of river water at Vejen Brook showed a concentration of 1 µg Cu/kg.

TABLE 6

COPPER ADDED (Cu, g/ha) TO SOIL FROM MANURE

Cu in manure DM, mg/kg	Cattle manure, t/ha with 27% DM			Pig manure, t/ha with 24% DM		
	50	100	200	50	100	200
30	405	810	1 620	360	720	1 440
60	810	1 620	3 240	720	1 440	2 880
90	1 215	2 430	4 860	1 080	2 160	4 320
250	3 375	6 750	13 500	3 000	6 000	12 000
500	6 750	13 500	27 000	6 000	12 000	24 000

Sludge from sewage plants often contains large amounts of copper. Table 7 shows the amounts of copper added to the soil by different amounts of sludge of two types, the first of domestic, the second of industrial origin.

TABLE 7

COPPER ADDED (Cu, g/ha) TO SOIL BY SEWAGE SLUDGE

Cu in sludge DM mg/kg	Household sludge, t/ha with 40% DM				Industrial sludge, t/ha with 25% DM			
	5	10	25	50	5	10	25	50
100	200	400	1 000	2 000	125	250	625	1 250
250	500	1 000	2 500	5 000	313	625	1 563	3 125
500	1 000	2 000	5 000	10 000	625	1 250	3 125	6 250
1 000	2 000	4 000	10 000	20 000	1 250	2 500	6 250	12 500
1 500	3 000	6 000	15 000	30 000	1 875	3 750	9 375	18 750
3 000	6 000	12 000	30 000	60 000	3 750	7 500	18 750	37 500

COPPER CONTENT OF CROPS

Table 8 shows the average Cu contents of various crops
(Damgaard-Larsen et al., 1979) differently manured in at 3-year
trial on three sites in Denmark. The differences in Cu content
are less than might be expected from the variation in Cu
application. It must be borne in mind that the industrial
sludge also contained large concentrations of other heavy metals
and the yields obtained were much lower than those from domestic
sludge.

SOIL Cu CONTENT AFTER 3 YEARS CONTINUOUS MANURING WITH SLUDGE

Table 9 shows the effects of the treatments in the above
experiment on available, and acid-soluble, soil Cu. The copper
applied had a large effect on the available Cu contents of
surface soil (0 - 25 cm) but very little effect in the subsoil
even when 134 kg Cu/ha was applied.

Acid soluble copper was determined in soil samples dried
at 50 - 60°C and then boiled with 7 N HNO_3 for 4 - 5 h. Cu in
the filtrate was measured by atomic absorption. Using this
'strong' method, more copper is extracted but, though the Cu

TABLE 8

CONTENT OF COPPER IN DIFFERENT CROPS WITH AND WITHOUT APPLICATION OF SLUDGE
TO THE SOIL, mg/kg IN DM

Crop	Control without Cu	Household sludge, 21 kg/Cu/ha	Industrial sludge, 134 kg/Cu/ha	Fertiliser without Cu
Barley, grain	4.0	4.8	4.9	3.6
Barley, straw	4.1	4.4	3.9	4.1
Oats, grain	3.6	4.3	3.9	3.1
Oats, straw	3.1	4.6	4.2	4.5
Grass	6.4	9.3	8.6	8.1
Beet, root	7.2	7.0	7.4	8.2
Beet, top	9.2	11.8	12.0	10.8
Potato	7.4	8.8	8.8	6.9
Kale, leaves	8.8	9.7	8.8	10.1
Kale, stalk	6.6	6.9	7.4	5.8
Carrot, root	9.3	10.1	9.4	11.0
Carrot, top	12.7	13.0	14.2	12.2
Cabbage	5.6	7.8	5.5	4.9

levels are much higher, the effect of Cu application is only
of the same order as that on available Cu content.

STANDARDS FOR COPPER APPLICATION

In order to provide guidance on the use of sewage sludge
in agricultural areas, several countries have published standards for maximum loads of heavy metals. Table 10 provides
estimates of safe upper limits of copper loading (Hucker, 1979).

In a report on pig slurry (Batey et al., 1972), it is
stated that up to 9.5 kg/Cu/ha can be applied for grass and
other crops without any harmful accumulation in the soil.

TABLE 9

CONTENT OF COPPER IN SOIL WITH AND WITHOUT SEWAGE SLUDGE, AVERAGE OF THREE SOILS

Soil depth, cm	Household sludge, 21 kg/Cu/ha	Industrial sludge, 134 kg/Cu/ha	Fertiliser without Cu
EDTA - Cu (= Cu_t), mg/kg			
0 - 25	3.0	7.9	1.5
25 - 50	0.8	1.2	0.7
50 - 75	0.7	1.0	0.8
75 - 100	1.0	0.9	0.9
Average 25 - 100	0.8	1.0	0.8
Acid-soluble Cu, mg/kg			
0 - 25	7.9	14.5	6.9
25 - 50	7.4	7.2	6.6
50 - 75	8.5	11.3	11.8
75 - 100	10.3	9.0	11.0
Average 25 - 100	8.7	9.2	9.8

TABLE 10

STANDARDS FOR MAXIMUM ALLOWED COPPER APPLICATION TO AGRICULTURAL SOIL IN DIFFERENT COUNTRIES

Country State or Province	Annual Cu supply, g/ha	Allowed period of years	Maximum allowed Cu supply, kg/ha
Ontario, Canada			165
England, general	9 300	30	280
England, grassland	18 600		560
Scotland	5 600	50	
The Netherlands	1 000	50	
Finland	12 000		
Sweden	3 000		
Wisconsin			364
Denmark[a]	4 500		

[a] Calculated on basis of proposal from the environmental board of Denmark concerning disposition of sewage sludge

CONCLUSIONS

The question of copper loading in agricultural areas has become topical in recent years as a result of the higher copper content of animal manures. This is due to the increased use of copper supplemented feeds, especially for pigs, and of copper solutions as 'foot-baths' for cattle. Cu-enriched fertilisers are also used.

Application of reasonable amounts of manure for agricultural crops may cause no risk of copper poisoning of crops or animals, though special care must be taken with sheep as they are Cu sensitive. At the time of writing, no case of copper poisoning of sheep has yet been reported in Denmark. Animals have been fed on grass or other crops which had been liberally treated with pig slurry, but care should be taken where copper-enriched pig slurry is used on grass intended for sheep. Slurry adhering to the grass leaves may increase the Cu intake above the safety level.

The question of copper loading in soils deserves attention and the effects should be evaluated by more long-term experiments and by observations on farms where copper use is high. In the long term, the grower of crops and the animal husbandry man have a common interest in this problem. It would be unwise to introduce so much copper to the farming system that crop Cu contents rise to unacceptably high levels.

REFERENCES

Andersen, C., 1978. Jordens kalk- og gødningstilstand. Ugeskr. Agron. 123, 971-977.

Barber, R.S., Braude, R. and Mitchell, K.G., 1955. High copper mineral mixture for fattening pigs. Chem. Ind. Vol. no. 601.

Batey, T., Berryman, C. and Linie, C., 1972. Disposal of copper enriched pig manure slurry on grassland. J. Br. Grassland Soc. 27, 139-43.

Damgaard-Larsen, S., Larsen, K.E. and Søndergaard Klausen, P., 1979. Yearly application of sewage sludge on agricultural land. Tidskr. planteavl. 83, 349-386.

De Haan, F.A.M., Lexmond, Th.M. and Dykman, F., 1976. Utilisation of manure by landspreading, pp 289-297. Luxembourg: Commission of the European Communities, EEC, EUR 5672 e.

Frederiksen, J., 1971. Hedeselskabets laboratorium. Beret Virksomheden 1970-71.

Hansen, V., Sunesen, N. and Bresson, S., 1974. Rapport 416 fra forsøgslaboratoriet. pp 4-23. Frederiksberg bogtrykkeri.

Hartmans, J., 1978. Animal and human health hazards associated with the utilisation of animal effluents, pp 35-56. Luxembourg: Commission of the European Communities, EEC, EUR 6009 EN.

Henriksen, Aa., 1957. Kobberbestemmelse i jord i sammenligning med virkning af kobbergødskning. 550, Beret Statens Forsøgsvirsomhed i Plantekultur, 685-717.

Hucker, G., 1979. Environmental effects of disposing sewage sludge on land. Symposium Cadarache, France. (in press).

Kjellerup, V. and Søndergaard Klausen, P., 1975. 1212. meddelelse fra Statens Forsøgsvirksomhed i Plantekultur.

Kjellerup, V. and Lindhard, J., 1977. 1313. meddelelse fra Statens Planteavlsforsøg.

Kjellerup, V. and Dam Kofoed, A., 1979. Nitrogen fertilising in relation to plant nutrients in drainage water. Tidsskr. Planteavl. 83, 330-48.

Lexmond, Th.M. and de Haan, F.A.M., 1977. Proc. Soil Environment and Fertility Management in Intensive Agriculture, pp 383-393. Tokyo: Society of the Science of Soil and Manure.

Nielsen, H.E., Madsen, A. and Just, A., 1979. Symp. European Association for Animal Production, pp 1-3.

GENERAL DISCUSSION

R.J. Unwin *(UK)*

I would just like to take up one point that Dr. Gasser made on Dr. Coppenet's paper. In terms of the total copper additions suggested in that paper, I do not think it is at variance with the type of view I was putting forward since Dr. Coppenet is referring to a range of agricultural crops which we would consider to be relatively resistant. He mentions a figure of 120 for EDTA copper as being somewhere about the critical value. For those crops I would have quoted a figure in excess of 100 as well.

I would like now to go back to the question of recovery of copper in soil. I would be interested to hear the experience of other people, with regard to the texture of the soil, to follow up the point made by Dr. McGrath. In the experiments to which I referred and where we had a recovery of copper in the order of 80%, the soil had a silt content of 45% and the clay content was 40%. So we had a soil which was about 90% silt + clay and there we had a very high recovery. I should add that in that same experiment, while we could account for recovery of copper, we could not account for recovery of zinc although we also knew exactly how much zinc had been applied.

D. McGrath *(Ireland)*

The situation for zinc may be slightly different from that for copper. In some cases with high levels of application of pig slurry containing a lot of zinc, we actually got a decrease in the zinc content of the soil. I should stress that this happened for sub-surface horizons and it indicated to me mobilisation of the metal by virtue of the high organic matter levels which were added.

A. Dam Kofoed *(Denmark)*

Is it possible to find other disinfectants rather than copper sulphate for loose housed cows?

J.H. Voorburg *(The Netherlands)*

Formaldehyde.

A. Dam Kofoed

Yes, we know about that but we do not like to use it.

K.L. Robinson *(UK)*

This discussion on the fate of copper in the soil and the inability in some cases to recover it, leads to the question of copper in surface and drainage waters. People concerned with the protection of the environment in various places are bandying about a lot of data on the toxicity of copper for fish and so on. The implication is that the application of pig slurry to land will lead to the contamination of surface waters. Is there any information available at this meeting on copper enrichment or drainage water from slurry applications, or copper in water from lysimeter experiments. I would be glad to hear any comments of that kind.

A. Dam Kofoed

As far as Denmark is concerned, under practical conditions we have a leaching of copper from 8 to 19 g/ha. It is roughly the same under lysimeter conditions. If we take samples from a small river and measure the copper there it will be 1 µg/l.

R.J. Unwin

To answer Dr. Robinson's question, I do not have any information with me from ADAS on MAFF experiments in England and Wales. I am not sure whether copper was determined in the leachates from the lysimeter experiments to which I referred; I would have to go back to the people who conducted that work. I would just add that studies have commenced with the application of sewage sludge in lysimeter experiments to sandy soil where copper is one of the residues which will be looked for. One should, perhaps, make reference to the information from Northern Ireland that Dr. McAllister has talked about quite often, where he has been able to show movement of whole slurry through drainage systems. So obviously there is a copper increase there, but also the pollution of the whole slurry getting into the system.

P.H.T. Beckett *(UK)*

We have measured the copper concentration of the soil solution; we have centrifuged the solution from the soil and then analysed it. It is unlikely that the effluent on drainage would be greater than this; it would probably be less. Even for exceptionally high sludge applications, equivalent to 200 years normal application, we find it is difficult to push the concentration above a very small number of ppm.

Th. M. Lexmond *(The Netherlands)*

I think it should be mentioned that copper in solution can be present largely in the form of complexed copper. I believe the toxicity of copper to aquatic organisms has also been shown to be related to the free cupric ion activity and not to total copper in solution. So, whenever results of this type are presented, it would be very useful to make the distinction between the complexed and the free cupric ion.

R. Braude *(UK)*

I know of a case in Yugoslavia where slurry from a piggery producing 100 000 pigs per year, with high copper supplementation, went straight into the river. According to the statement they gave, it had no effect on fish - and there were fish in that river.

G.A. Fleming *(Ireland)*

I would like to make a comment regarding any attempt that might be made at estimating recoveries of added copper to different soils. I would take you back to some work that Dr. le Riche did at Rothamsted in the 1950's, where he showed that a number of trace elements were bound at varying degrees of tightness to the iron oxides and manganese oxides in soils. A number of trace elements are associated. We know that elements such as cobalt and lead are very tightly bound with manganese oxide. Zinc and copper can be bound to some extent but they seem to be more easily leachable. The point I want to make is that when looking at different soils it would be foolish to ignore the possible effects of oxides of iron, and manganese, because they occur in so many different forms that they can bind trace metals with varying degrees of tenacity.

<u>Th. M. Lexmond</u>

I think you can distinguish between two mechanisms binding heavy metals to manganese and iron oxides. One of the mechanisms is surface adsorption. One would expect copper, when surface adsorbed, to be dilutable by, for example, diluted nitric acid. Another mechanism which may account for binding in the longer term, may be occlusion within the crystal of the oxides. That copper would be impossible to recover with a short extraction with dilute nitric acid, or EDTA.

<u>P.H.T. Beckett</u>

Dr. Lexmond's Figure 3 plots copper activity against pH. This is not, of course, a desorption isotherm, but it does suggest that for the groups of soils he presents, the free energy of copper and hydrous oxide are a constant sum and we might suppose that we have copper and hydrogen antagonism at his sites. I think that is the correct inference from that diagram.

<u>Th. M. Lexmond</u>

I agree with that but I do not agree with your remark that it is not a desorption isotherm. It surely indicates an antagonism between protons and cupric ions in the binding of copper in soils, particularly in this soil, of course.

P.H.T Beckett

Well, it is a plot of two free energy terms, one against the other, whereas a desorption isotherm would be a quantity against free energy.

Th. M. Lexmond

Yes, OK, it has been derived from a desorption isotherm.

P.H.T. Beckett

With regard to Figure 1, whether we disagree or not I do not know, but perhaps over lunch we can look at the yield data to see whether we disagree.

Th. M. Lexmond

The yield is not much affected in this stage of growth, but an important parameter might be the organic nitrogen content of the shoots which decreases when copper becomes toxic to the crop.

P.H.T. Beckett

Well, if the yield is not reduced there is no toxicity, so we have no argument.

Th. M. Lexmond

I think you saw the slides on morphology of the roots and I think it is quite difficult to maintain that there is no toxicity.

P.H.T. Beckett

The farmer is only concerned with the portion that is harvested.

Th. M. Lexmond

I think the experimental technique used may lead to different results. When you grow plants for a few days on solution cultures there is no apparent effect on yields at the beginning but it shows up when you continue to grow your plants for say 14 or 20 days.

J.H. Voorburg

I have one final comment before closing this session. Copper can also escape from the soil in a different way. Copper in soil has an influence on worms; the worms have a higher copper content and they are caught by birds. I do not know whether the birds grow better on it but there is an influence on the environment.

COPPER IN SLUDGE - ARE THE TOXIC EFFECTS OF COPPER AND OTHER HEAVY METALS ADDITIVE?

P. Beckett

Department of Agricultural Science, University of Oxford,
Parks Road, Oxford OX1 3PF, UK.

ABSTRACT

National regulations on the disposal of sewage sludge onto agricultural land differ considerably in their assumptions about the additivity of the toxic effects of Cu, Ni or Zn. This paper demonstrates firstly that the nature of the assumptions made greatly affects how much of a given sludge may be disposed and then presents evidence in support of a middle position.

THE PROBLEM

There is concern about the problems of disposing sewage
sludge onto agricultural land when it contains Cu, Ni and Zn.
If too much sludge is applied the land is poisoned. If sludge
may not be disposed in this way it must be dumped or
incinerated. These methods are considerably more expensive
and create severe pollution problems of other kinds. They
also break the chain of recycling natural resources. Water
and sewage authorities have the difficult problem of finding
a safe and acceptable path between these unacceptable
alternatives.

The legislating organisations that control sludge
disposal have very little real information on which to base
regulations. The general ignorance is particularly con-
spicuous on the question of whether the toxic effects of Cu,
Ni and Zn are additive or not. Thus draft regulations of the
Environmental Protection Agency of the USA assume their
effects are not additive; draft regulations for the UK assume
that they are. (See Lewin and Beckett, 1980).

The disagreement sounds small, just a matter of wording,
but it has rather large implications. Consider a typical
urban sludge that contains 1 167 mg Zn, 583 mg Ca and 146 mg
Ni per kg dry solids. In the USA a quantity of this sludge
equivalent to 428 tonnes dry solids/hectare may be disposed
onto a loamy soil, pH $\geqslant$ 6.5, that has not received sludge
before; in the UK only 160 tonnes/hectare. The two countries
disagree on whether the relative toxicities of Zn and Ni are
in the ratio 1 : 5 or 1 : 8, but this affects the contrast by
only 15%. The difference is almost wholly due to different
assumptions about additivity. One or other country would
seem to be making an expensive mistake.

This paper examines further the practical implications
of additivity (or otherwise) on sludge disposal, and how they
are affected by different degrees of additivity. It then

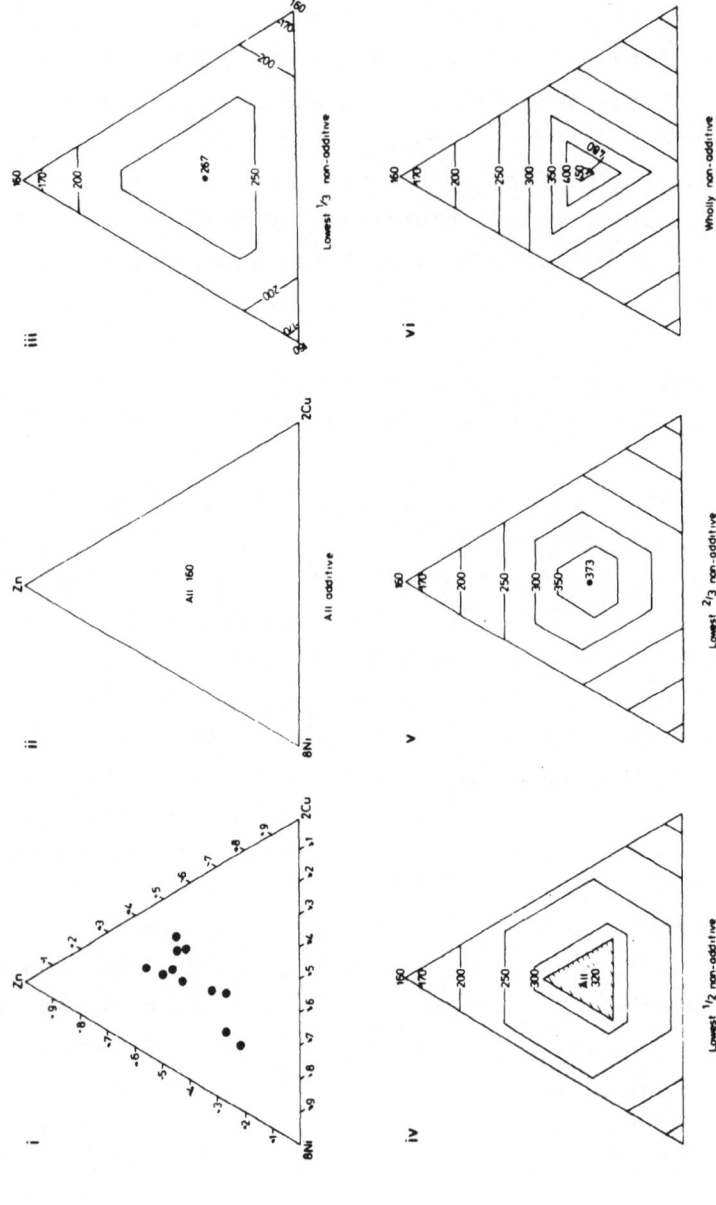

Fig. 1: Maximum cumulative (30 year) sludge applications in tonnes dry matter/ha, for sludge of varied proportions of Cu, Ni and Zn but constant (Zn + 2Cu + 8Ni) = 3500 mg/kg sludge dry solids, and different assumptions about the additivity of their toxic effects (see text). are the 11 principal sludges from a major UK city.

reports briefly some preliminary trials to measure the extent
of additivity.

THE PRACTICAL RESULTS OF DIFFERING DEGREES OF ADDITIVITY

In order to examine these implications further I have
tentatively accepted the UK conventions that:
(i) The relative toxicities of Zn, Cu and Ni are in the ratio
     1 : 2 : 8, so that the total heavy metal 'load' of a
     sludge is given by its Zinc Equivalent or (Zn + 2 Cu +
     8 Ni) in mg/kg sludge dry solids.
(ii) the maximum cumulative application of sludge over 30
     years is:
$$T = \frac{5.6 \times 10^5}{\text{Zinc equivalent}} \quad \text{tonnes dry solids/hectare.} \quad (1)$$

The exact values of the constant terms have little effect
on the discussion that follows.

Then I have created a range of hypothetical sludges by
varying the proportions of Cu, Ni and Zn in the typical sludge
above while keeping their total load constant (i.e.
Zn + 2 Cu + 8 Ni = 3 500 mg/kg DS). The different combinations
are represented on triangular diagrams like Figure 1 (i). This
also shows the compositions of the principal sludges produced
by a major city in the UK.

The rest of Figure 1 indicates the maximum cumulative
application of each sludge that may be disposed (tonnes DS/ha
in 30 years), on various assumptions about the additivity of
the three elements. Figure 1 (ii) assumes total additivity
as in the UK regulations. Figure 1 (vi) assumes the three
elements are totally independent (no additivity) as in the US
regulations. Figures 1 (iii) - 1 (v) assume partial additivity,
i.e. that of the total amount of an element added to land in
sludge, only that part in excess of 1/3, ½ or 2/3 of the
quantity of it that would be toxic were it the only heavy
metal present, contributes to the total toxic load. The

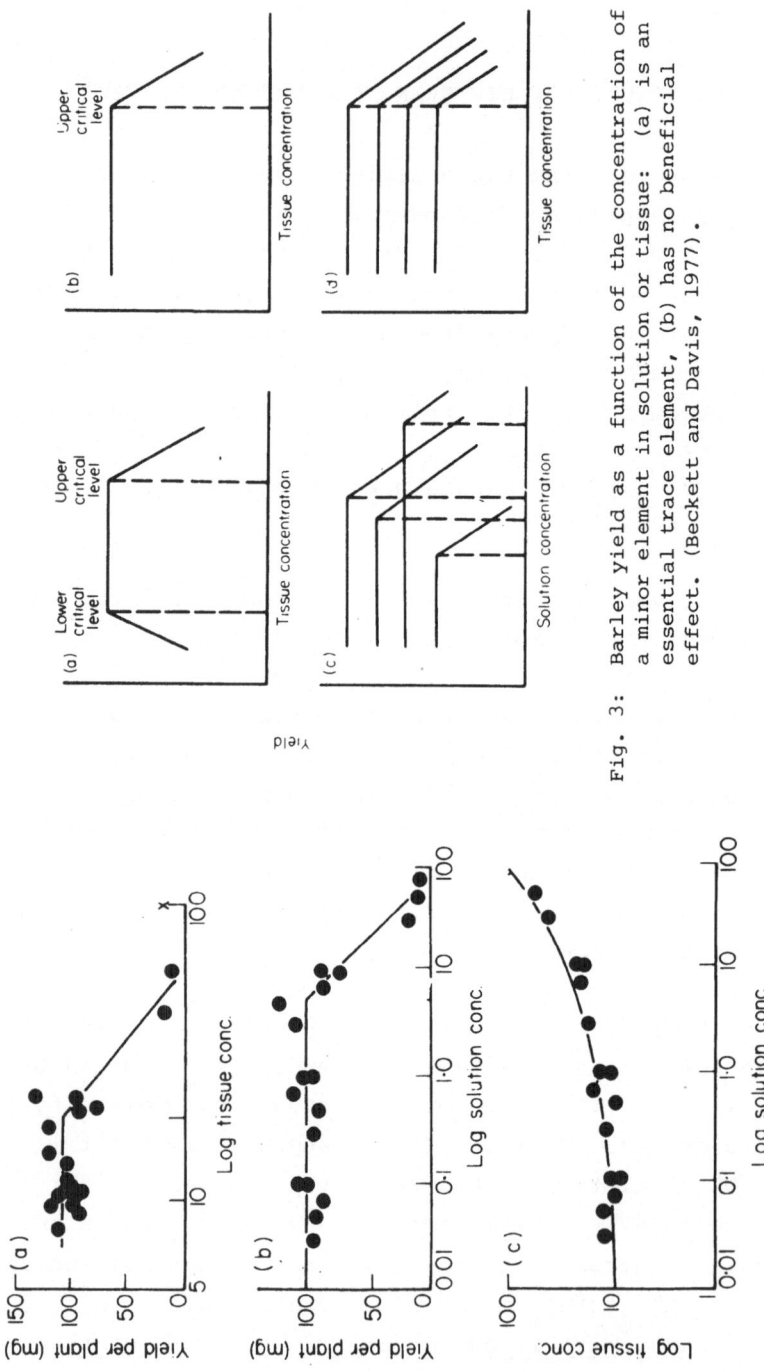

Fig. 3: Barley yield as a function of the concentration of a minor element in solution or tissue: (a) is an essential trace element, (b) has no beneficial effect. (Beckett and Davis, 1977).

Fig. 2: Copper uptake and toxicity in young barley (concentrations in mg/kg dry matter or mg/l solution) (Davis and Beckett, 1978).

calculations show clearly that for a sludge containing roughly
equivalent quantities of the three elements, as many do
(Figure 1 (i)), the quantity that may be disposed is very
sensitive to the extent that the toxic effects of its
constituents are additive. Unfortunately there is not very
much real information on this for field crops and much of this
relates to sites that are already very heavily polluted. Also
much of it fails to distinguish:

> the effect of one element on the uptake of a second
> element and its translocation to the leaves or edible
> portions of the crop;

from the effect of one element in a leaf or other organ on
> the toxicity of a second element also present in it.

SOME EVIDENCE ON ADDITIVITY

For a slightly different purpose, Dr. Davis and I have
examined the simultaneous effects of Cu, Ni and Zn on barley
(Beckett and Davis, 1977, 1978; Davis and Beckett, 1978).
We worked with young barley at the onset of tillering. This
is a sensitive stage. The crop was grown in sand, which was
frequently irrigated with nutrient solutions containing varied
proportions of Cu, Ni and Zn, and harvested at the five-leaf
stage. Yield was measured as mean dry weight per plant, and
heavy metal uptake in terms of their concentration in the
above-ground dry matter.

For any one of these elements alone uptake was
approximately proportional to its concentration in the
nutrient solution, at low concentrations, but rose more sharply
at high concentrations (Figure 2c). The form of this relation
varied considerably with season and growing conditions. Graphs
of yield against Log (tissue concentration) or Log (solution
concentration) showed sharp inflexions at the onset of toxicity
(Figure 2a, b) and then a linear decrease in yield in response
to further increases in the toxic element. The critical
solution concentration varied considerably with season and

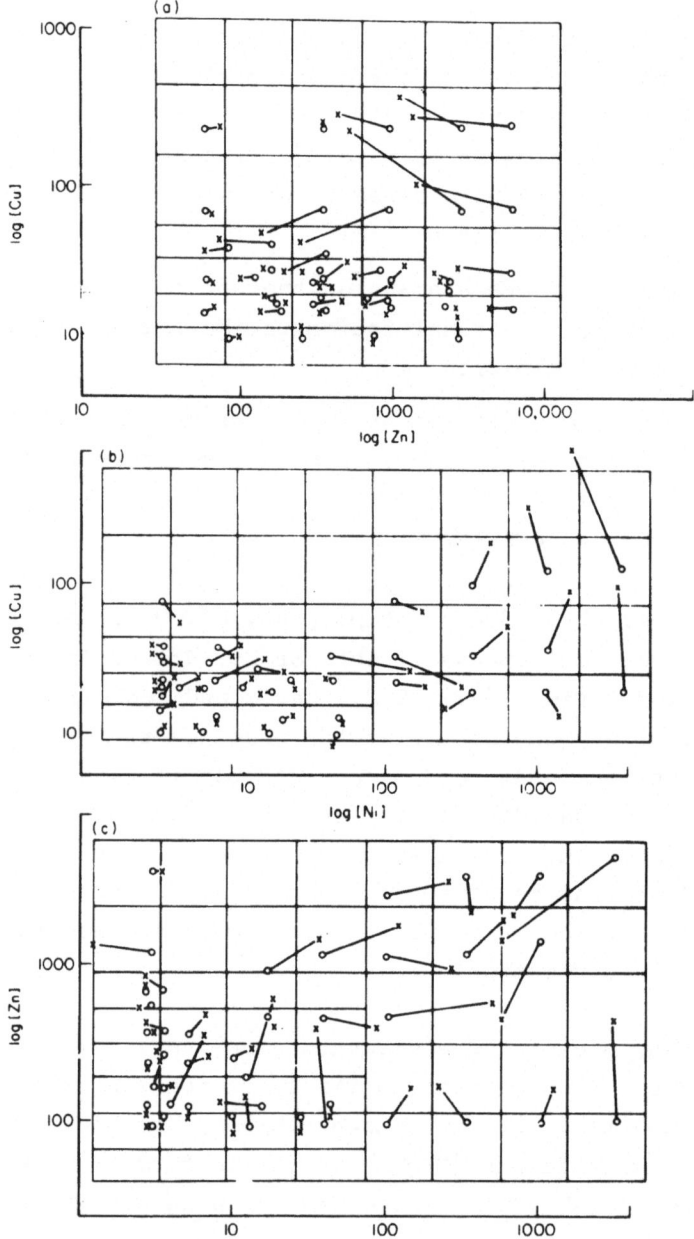

Fig. 4: Tissue concentrations of young barley grown in nutrient solutions with varied concentrations of Cu and Zn, Cu and Ni, Ni and Zn. X are the tissue concentrations found; O are the concentrations expected had the other element not been present. (Beckett and Davis, 1978).

growing conditions. Graphs of yield against Log (tissue concentration) or Log (solution concentration) showed sharp inflexions at the onset of toxicity (Figure 2a, b) and then a linear decrease in yield in response to further increases in the toxic element. The critical solution concentration varied considerably with season and growing conditions, while the critical tissue concentration was relatively independent of these (Figure 3). The critical tissue concentrations of Cu, Ni or Zn in young vegetative tissues were relatively constant for different crop species.

Effects on uptake

Figure 4 presents the results of three experiments with nutrient solutions containing two heavy metals. For each treatment in each experiment it plots the barley tissue concentrations that would have been expected (o) if the second element had not been present, and also the tissue concentrations actually found (X). The expected and real values for each treatment are linked by a line.

Wollan (1977) has shown that even in a field that had received 6 times the permitted cumulative application of a sludge (i.e. 240 times the normal annual application) the concentrations of soluble Cu, Ni and Zn in its soil solution do not exceed 1.5, 0.5 and 1.5 mg/l respectively. In the experiments of Figure 4 these correspond to tissue concentrations of Cu 10 - 20, Ni 7 - 15, Zn 100 - 200 mg/kg plant dry matter. Clearly the three elements have very little effect on the uptake of each other at the range of concentrations we are likely to meet.

Effects on toxicity

For one of the same experiments, Figure 5 plots crop yields against tissue concentration. Clearly there is a range of non-toxic combinations of the two elements for which yield is unaffected: 95 ± 15 mg/plant ('yield plateau'). Then there is a relatively rapid fall in yield as their concentrations increase beyond the limits of the plateau

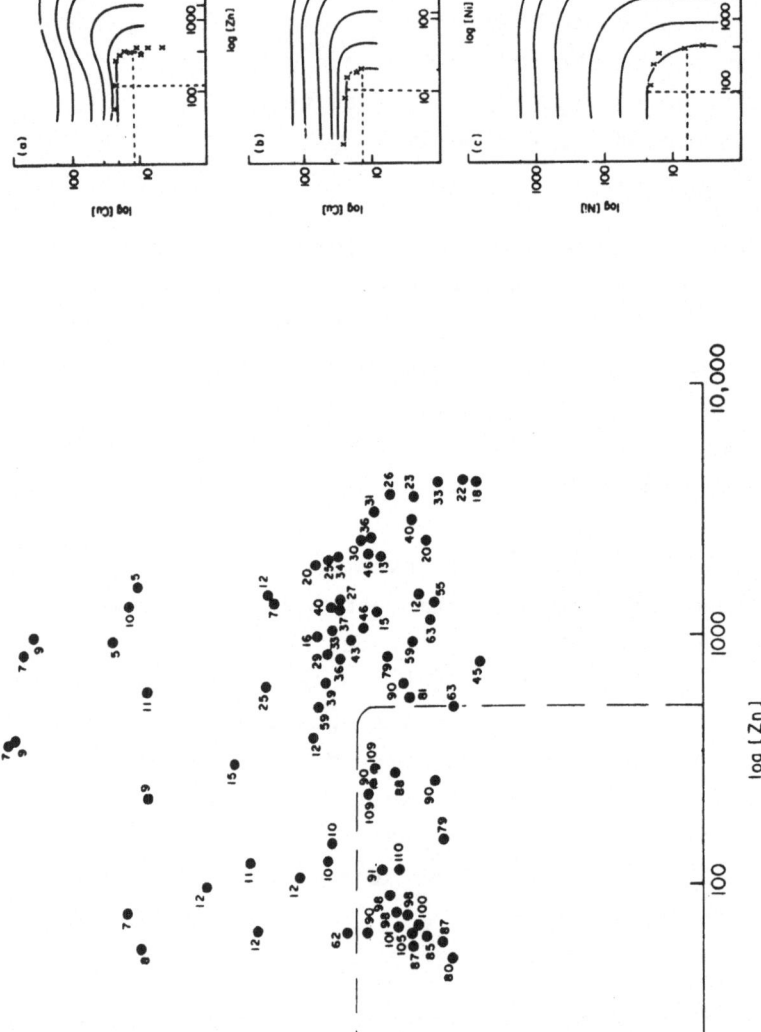

Fig. 5: Barley yields (mg dry matter/plant) from the first experiment of Fig. 4. - - represents the plateau edge (non-toxic limits) of Fig. 6. (Beckett and Davis, 1978).

Fig. 6: Yield contours for 100%, 75%, 50%. 25%, 10%, 0% of plateau yield constructed from Fig. 5 and similar results for Ni/Cu, Ni/Zn X marks the plateau edge. (Beckett and Davis, 1976).

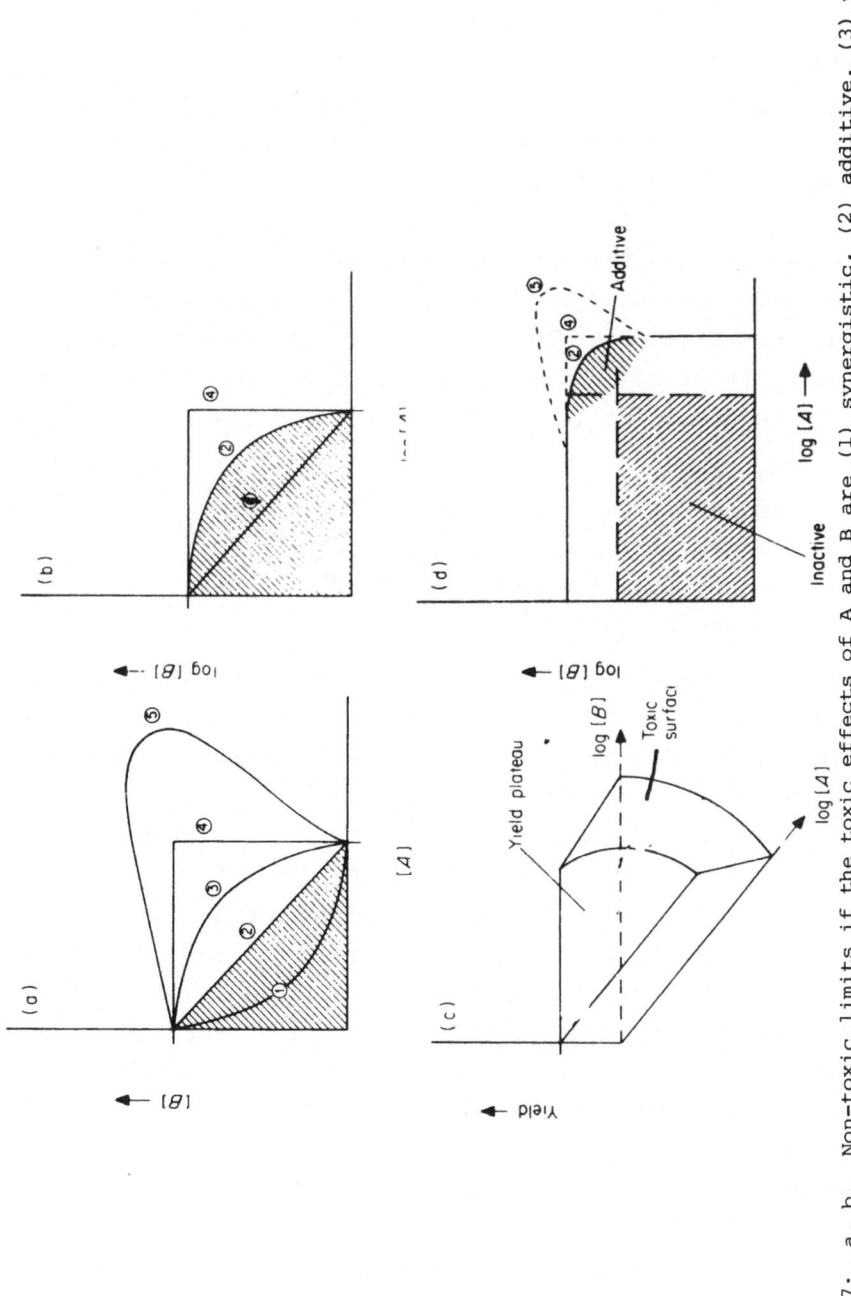

Fig. 7: a, b. Non-toxic limits if the toxic effects of A and B are (1) synergistic, (2) additive, (3) weakly additive, (4) independent (non-additive) and (5) antagonistic.
c, Response surface to elements A and B.
d, Model of interaction that fits the data of Fig. 6 (Beckett and Davis, 1978).

('non-toxic limits'). Figure 6 plots contours of diminishing
yield. The original paper (Beckett and Davis, 1978)
illustrates how they were derived from the data. These contours
describe three-dimensional figures like Figure 7c. Note
particularly that a high proportion of these contours is
parallel to one or other axis. Where this is so additions of
a second element are not altering the concentration of the
first element necessary to cause toxicity.

When we examine the various kinds of interactive effects
that two toxic elements can display, they are all rather
different from Figure 6. Thus Figures 7a and 7b illustrate
(arithmetically or logarithmically) the non-toxic limits that
must be expected if the toxic effects of elements A and B were

1)   synergistic
2)   additive (UK model)
3)   weakly additive
4)   independent (US model)
5)   antagonistic.

On comparing these with Figure 6 it is difficult to avoid the
inference that the greater part of the Cu, Ni or Zn in plant
tissue (up to ½ - 2/3 of their concentrations at the non-toxic
limit) is not only inactive but wholly non-additive. It is
only the quantities of them in excess of ½ to 2/3 of their
critical levels that are additive (Figure 7d). If so, and
since the relation between tissue concentration and solution
uptake is roughly linear in the range that concerns us,
Figure 1 (iv) may provide a better basis for regulations than
Figures 1 (ii) and 1 (vi). Incidentally the effects of Cu and
Zn appear to be antagonistic - they may be less toxic in the
presence of each other than alone.

CONCLUSIONS

1.   The amount of sewage sludge that may be safely disposed
      onto agricultural land depends very much on whether the
      toxic effects of the Cu, Ni or Zn in it are additive or

not. This being so, it is important to discover whether, and how much, their effects are additive.

2.   Preliminary results suggest that these three elements show less additivity than is assumed in the draft UK regulations, but a little more than is assumed in the draft US regulations. If generally applicable this would enable many UK authorities to double their sludge application rates.

REFERENCES

Beckett, P.H.T. and Davis, R.D., 1977. Upper critical levels of toxic elements in plants. New Phytol., 79, 95-106.
Beckett, P.H.T. and Davis, R.D., 1978. The additivity of the toxic effects of Cu, Ni and Zn in young barley. New Phytol. 81, 155-173.
Davis, R.D. and Beckett, P.H.T., 1978. Upper critical levels of toxic elements in plants, II Critical levels of Cu in young barley, wheat, rape, lettuce and rye grass, and of Ni and Zn in young barley and rye grass. New Phytol. 80, 23-32.
Lewin, V.H. and Beckett, P.H.T., 1980. Monitoring heavy metal accumulation in agricultural soils treated with sewage sludge. Effluent & Wat. Ttmt. J. 217-221.
Wollan, E. 1977. Availability to plants of heavy metals in soils treated with sewage sludge. M.Sc. thesis, Oxford University.

DISCUSSION

J.H. Voorburg *(Netherlands)*

I remember Mr. Williams from Northampton has a zinc
index equivalent'. He says that copper and zinc are additive.
That is a different opinion from our country.

P.H.T. Beckett *(UK)*

Sometimes it is difficult to demonstrate imperfections in
additivity from experimental data. For example, Dr. Coppenet's
data would appear to indicate additivity unless you keep an
open mind on the possibility of an alternative interpretation.

J.K.R. Gasser *(UK)*

You have dealt with two elements here, and we have been
discussing the question of three or more. Mathematically,
have you developed the model to enable you to deal with the
questions of independence, additivity, synergism and antagonism,
using three or more elements, because this requires multi-
dimensional mathematical models?

P.H.T. Beckett

We have done the calculations but we have not made
drawings (in four dimensions) of three elements. For copper,
nickel and zinc jointly, if you take their proportions from
experiments with the elements in pairs, and assume that only
the amount of each element above two-thirds of their critical
level is additive, you can then apply this to the three
element situation and you get the right answer. That is the
most that we have done.

M.D. Webber *(Canada)*

Assuming that metals are added to soil in sludge, is the
additive effect a continuing effect? Would it be observed
over a period of several years, as opposed to just within the
short period of time when the sludge was first added?

P.H.T. Beckett

Remember that these are tissue concentrations. The difference between one crop and another, or one year and another, is that different amounts are taken up. The lettuce is a sensitive crop, not because it is particularly vulnerable when the elements are in the leaf but because it takes up more from the same soil than barley does. Our results would apply equally to the second or third year because they relate to the heavy metals in the tissue and not in the soil. You may now ask how do you relate tissue concentration to soil concentration and I cannot answer that at the moment.

P. Worthington *(UK)*

If the guidelines for the disposal of sludge to land are designed properly, then the effect on the plant, whether it is synergistic or additive, will not be noticed because the dosing of the metals will be stopped before this point is reached. Therefore, this problem that you have raised should never arise in practice.

P.H.T. Beckett

True as far as you go. The problem arises very brutally since the calculations of when you will reach the stopping point lead to a completely different answer if the elements are additive, or if they are not. For a sludge which is median to all the sludges of one of our major cities 160 t/ha is the limit. According to the current UK regulations, if the effects of Cu, N and Zn were non-additive, you could go up to 380 t/ha before you would reach toxicity.

R.J. Unwin *(UK)*

Let us not confuse this discussion by reference to the current guidelines and regulations for sludge disposal in the UK. Really this is to emphasise the Chairman's comment about the possible discrepancy between Mr. Williams' comments and what is in the guidelines. Those original suggestions were

made with an acknowledgement that we did not have the type of
information that Dr. Beckett is now providing with his research.
This is the type of information we have been looking for for
the last ten years and, as he says himself, it still has a way
to go before we can relate soil concentrations to tissue concentrations. I don't think there is any great dispute between
what we have as guidelines and what Dr. Beckett has been
putting to us today.

A. Dam Kofoed (Denmark)

To what extent has this additive effect been taken into
account in relation to the stated English limit of 9.3 kg/ha
for copper?

P.H.T. Beckett

I think one of my colleagues should answer this one.

R.D. Davis (UK)

As I interpret the present guidelines, the limit they
impose is not necessarily 9.3 kg/ha of copper. It depends on
how much nickel and zinc there is in the sludge. So the figure
floats according to the assumptions on additivity that are
implicit in the zinc equivalents of Mr. Williams. If there was
a lot of zinc or a lot of nickel in the sludge, the limit might
be reached with no copper.

P.H.T. Beckett

The answer then is that the present guidelines assume
complete additivity.

T.W.G. Hucker (UK)

I think the point really is that you could never put on
a level of 9.3 kg/ha because you have always got to reduce it
because there is always some zinc and some nickel. The UK
guideline is the only one I can think of which does take into
account the possibility of additivity. The rest of the world

seems to have zinc, copper, nickel as separate criteria. What
sort of modification to the current zinc, copper and nickel
criteria, would Dr. Beckett suggest to make life simple for us
in the British Isles? Would his investigation indicate a
reduction in the order of 10 or 15%, for example?

P.H.T. Beckett

In Britain we are, perhaps, a little illogical because
we only assume additivity of zinc, copper and nickel, whereas
copper and molybdenum, for example, which are much more closely
related, would be a more logical pair. What I am saying is
that if you have a sludge which is proportionately high in zinc
and moderate in copper and nickel, I believe it would be safe
to apply more sludge than we do at the moment because when we
reach a toxic level for zinc alone we are still far short of
the copper and nickel toxic levels, and short of the levels at
which there is any question of additivity. If we assume complete additivity then we are including in our calculations
of toxicity elements which are present but which are not
present in toxic amounts. The US guidelines make the calculation separately for each element and then plan the sludge
disposal on the one for which toxicity is achieved first. I
think they may be a little too optimistic for a sludge where
the three elements are in equivalent amounts : let us say that
they have a sludge with high zinc, fairly high copper and
fairly high nickel, then when they reach the toxic levels for
the zinc, the copper and nickel are already a little toxic and
should have been taken account of. In Britain we are including
second and third elements even when they are not toxic. I
believe that somewhere between these extremes there might be a
safe alternative.

G.A. Fleming (Ireland)

I don't know whether anyone has read Dr. Purves' book on
'Trace Element Contamination of the Environment'. He suggests
that cadmium might be used as an index for the amount of sludge
that it may be desirable to put on.

J.H. Voorburg

Yes, but will chemical analysis for cadmium show that there are residues of PCB, for example? One has to be very careful with sludges from waste water that has high amounts of zinc, cadmium, PCB, dieldrin, and so on; they should not be spread on arable land. So it is not sufficient to take account only of zinc and copper, or cadmium and so on. There are many other contaminants.

Th.M. Lexmond *(Netherlands)*

I would like to stress that there is one level at which the interaction phenomena can be observed - the plant level. On the other hand, there is the possibility of an interaction at the soil level. The simultaneous addition of several metals might lead to competition for a limited number of binding sites and thereby lead to a smaller increase in activity than would be observed if the metals were added individually.

P.H.T. Beckett

This is true. However, it would be very interesting to discover how far what you and I both think of as 'sites' in the soil are really sites for ion exchange. Canadian work, for example, would suggest that zinc in the soil solution may be controlled by solubility products and not by sites. Copper probably is controlled by sites on organic matter, but not all of it. You may be right if they are competing for the same sites, but if they are controlled by different insoluble salts, then they are not competing. I don't know the answer but I wish we did.

Th.M. Lexmond

I have only some experience which was gained in an experiment in which copper and nickel were added individually and together. Both nickel and copper activity was higher with the combined applications than with the individual additions. It certainly could not have been an effect of pH because pH

was not affected at all. Consequently, the effect on yields of
maize was much more pronounced in the combined applications
than in the individual applications.

P.H.T. Beckett

What was the texture of the soil?

Th.M. Lexmond

It was a sandy soil with organic matter as the main metal
binding constituent.

P.H.T. Beckett

I don't know the answer but I could prevaricate by
suggesting that some of the mechanisms I mentioned are going to
be less significant on such a soil.

I. Bremner (UK)

If we draw analogies with what happens in the animal
kingdom, I might suggest that if you are looking for a constant
type of interaction between metals such as copper and zinc,
then perhaps you are oversimplifying the situation. In the
animal kingdom, if you are looking at the toxicity of cadmium
or zinc, you can get rather complex interactions taking place.
If you happen to deal with a situation where the intake of
copper is limited then you can find that some of the toxic
effects of both cadmium and zinc will be directed towards production of a copper deficiency syndrome. So, in this circum-
stance, you find that the cadmium and zinc act together and
tend to exacerbate the copper deficiency. If, on the other
hand, you have an animal with a copper status which is quite
adequate, then you may find that the effect of cadmium is
directed more towards the metabolism in the kidney - you get
the classical renal damage. In these circumstances you find
that the zinc is actually quite beneficial because it tends to
inhibit the accumulation of cadmium. So, depending on the
overall balance between these different metals, you find that

cadmium and zinc may work together or they may work against each other.

P.H.T. Beckett

When we began this work we consulted textbooks on toxicology, most of which appeared to relate to fish. Certainly with fish swimming in waters of these compositions, the effects of the elements are additive. However, I think the feeling was that they were all operating at a rather similar point in some rather crucial pathway. The generation of dry matter by photosynthesis is probably a fairly complicated process and the toxic elements we are considering here all hit it at different places. To this extent they may be synergistic, but they don't have to be. Dr. Lexmond has been rather forbearing because he could have pointed out, as he did this morning, that these elements can, in fact, reduce or increase the uptake of macronutrients, such as phosphorus, which will produce further effects on dry matter production. In extreme cases of very high toxic elements or very low P my simple picture will be confused because the toxic elements are having a phosphorus effect and not, in fact, a direct effect on photosynthesis.

COPPER UPTAKE FROM SOIL TREATED WITH SEWAGE SLUDGE AND ITS IMPLICATIONS FOR PLANT AND ANIMAL HEALTH

R.D. Davis

Soil Science Section, Water Research Centre,
Elder Way, Stevenage, Hertfordshire SG1 1TH, UK.

ABSTRACT

Results of field and pot trials are reported which have examined the transfer into crops of copper added to soil in sludge. A pot trial with 39 different crops indicated which species are most sensitive to soil Cu and showed that plant genotype is an important factor determining the availability of Cu added to soil in sludge. In a field trial, less than 0.1% of Cu added to soil in sludge was recovered in the standing crop. An addition of 443 kg/ha of Cu in sludge increased the Cu concentration of herbage by 2.2 mg/kg to 6.8 mg/kg. In comparison, the phytotoxic threshold concentration of Cu in plant tissue is approximately 20 mg/kg and a level of 10 mg Cu/kg is required in the diet of ruminants. Uptake of Cu was greatest in the first year after application of sludge and declined thereafter. Significantly higher concentrations of Cu were found in ryegrass grown in pots compared with field plots which had received an equivalent amount of sludge. Results of pot trials must be interpreted with caution. It is concluded that Cu uptake from soil treated with sludge at rates associated with normal operational practice is unlikely to lead to deleterious effects on crop yield or the health of farm animals.

INTRODUCTION

During sewage treatment approximately 90% of the copper
in crude sewage is separated into sludge (Lester et al., 1979).
As a result, Cu is one of the metals which occurs in significant concentrations in sewage sludge. UK sludges commonly
contain about 600 mg/kg dry solids (d.s.) of Cu but a range of
Cu concentrations in sludge of 200 - 8 000 mg/kg d.s. has been
reported (Berrow and Webber, 1971). Sewage of entirely resid-
ential origin has a Cu concentration of about 0.08 mg/l which
produces a sludge containing 230 mg/kg d.s. (Davis and Carlton-
Smith, 1980a). The concentration of Cu in sludges from works
with predominantly domestic catchments is a little higher at
about 400 mg/kg d.s., associated with a Cu concentration in
sewage of about 0.2 mg/l (Matthews, 1980). Concentrations of
Cu in sludge in excess of this level are the result of industrial activity. For instance, at a large sewage treatment
works near London, where a limit of 5 mg/l of Cu in industrial
discharges to sewers is applied, the concentration of Cu in the
sludge was about 950 mg/kg d.s. in 1978/79 (Thompson, 1979).

Utilisation as a fertiliser of agricultural land is the
principal disposal route for sludge in the UK and accounts for
about 50% of the 30 million wet tonnes (1.25 million tonnes
d.s.) of sludge produced annually. In contrast to sludge, the
background concentration of Cu in most soils is only about 20
mg/kg with a range from 2 - 100 mg/kg (Berrow and Webber, 1971).
Whilst Cu and other heavy metals continue to occur in higher
concentrations in sludges than in soils it will be necessary to
control applications of sludge to agricultural land to avoid
possible toxicity problems. If sludge is applied to soils with
a pH value of at least 6, which is the usual practice in the UK,
leaching of most metals, including Cu, is likely to be minimal.
Instead, the metals accumulate in the top soil and any hazard
relates to their availability for plant uptake on which phytotoxicity and entry into the foodchain depend. This paper
reports on three experiments designed to assess the transfer
into crops of Cu added to soil in sewage sludge, and discusses

the implications of the results for plant and animal health.

MATERIALS AND METHODS

Trial 1

39 varieties of crops were grown in polythene pots, each
holding 10 kg of soil, in a glasshouse. The soil (loam, pH
value 6.7, and Cu content 113 mg/kg) was obtained from a site
with a long history of sludge deposition made on a sacrificial
basis. Leaves (green tissue) and edible parts of crops were
harvested at economic maturity except for those crops where the
leaves would be moribund at that stage (cereals, sunflower,
rape). Leaves of cereals were harvested at the tillering stage;
maize, sweetcorn and rape leaves were sampled during flowering.
Ryegrass leaves were cut at 20 - 25 cm in length. Storage
roots and tubers were washed with water; and potatoes, swedes,
turnip, sugar beet and carrot were peeled. A detailed account
of this experiment is given by Davis and Carlton-Smith, 1980b.

Trial 2

This was a field trial investigating the use of sewage
sludge in land reclamation, in which large amounts of air dried
digested sludge (Cu content 1 040 mg/kg d.s.) were applied to
a site originally excavated for gravel and subsequently filled
with urban refuse and capped with clay (pH value 7.9, Cu content 39.8 mg/kg). Sludge was applied in the autumn of 1976 at
four rates to plots of 20 m x 2 m and cultivated into the soil
to a depth of 20 cm. Details of the rates of application are
shown in Table 1. Each treatment was replicated six times.
The plots were sown with a mixture of fodder oats and perennial
ryegrass and two cuts of herbage were taken in each of the
following three summers. Yields of herbage were measured and
samples were retained for analysis.

Trial 3

This trial compared concentrations of Cu in ryegrass
(*Lolium perenne* L. cv. S24) grown on soil treated with equivalent

quantities of sludge applied either to pots, each holding 9 kg
of soil, or to field plots of 8 m x 1.3 m. The soil used was
a sandy loam (pH 6.1, Cu content 15.5 mg/kg), and the digested,
lagoon-matured sludge contained 980 mg Cu/kg d.s.; Table 1
presents the application rates of sludge. Each treatment was
replicated. In the field investigation, the application rates
shown in Table 1 comprised two dressings of sludge made in
1978 and 1979. The first dressing was rotovated into the top
15 cm of soil before sowing ryegrass and the second dressing
was applied to the surface of the plots. Although grass was
also cut in 1978, the results presented for this trial relate
to the first cut of ryegrass taken in the summer of 1979. In
the pot trial a single application was made in early 1979 and
mixed throughout the soil before sowing ryegrass in the spring.
One set of pots was left outside, and the second set transferred
to a glasshouse. The pot trial results shown relate to the
first cut of ryegrass taken in 1979 when the leaves were 20 - 25
cm in length.

In all three trials soil samples were air dried and
ground to pass 0.5 mm before extraction with concentrated nitric
acid and determination of Cu by atomic absorption spectroscopy
(AAS) using a Perkin Elmer 460 instrument. Plant material was
dried at $80^{o}C$ in a draught oven, ground to pass 2 mm and
digested with nitric acid before determination of Cu by AAS.

TABLE 1

RATES OF APPLICATION OF SLUDGE IN TRIALS 2 AND 3

| | Application rate | Treatment | | | | |
		S_1	S_2	S_3	S_4	S_5
Trial 2	Sludge dry solids (t/ha)	0	53	107	213	426
	Sludge Cu (kg/ha)	0	55	111	222	443
Trial 3	Sludge dry solids (t/ha)	0	5.4	10.8	16.2	-
	Sludge Cu (kg/ha)	0	5.3	10.6	15.9	-

RESULTS

Trial 1

Figure 1 shows the results for the 39 varieties of crops, arranged in order of increasing concentrations of Cu in the leaves. Concentrations of Cu in the leaves always exceeded concentrations in non-foliar edible parts and for most crops fell within the range 5 - 15 mg/kg dry matter (DM) normally observed for Cu in plant tissue. The lowest concentration was seen in onion. Highest concentrations occurred in leaves of members of the beet family (*Chenopodiaceae*) such as sugar-beet, mangold and beetroot, and high concentrations also occurred in cereal leaves.

Trial 2

Figure 2 shows that increases in soil concentrations of Cu were a linear function of Cu added in sludge; the correlation coefficient (r value) associated with the results was highly significant (P = < 0.001). The highest rate of application of sludge of 426 t/ha of dry solids added 443 kg Cu/ha to the soil (Table 1) and increased the concentration of Cu in the soil by 125 mg/kg. However, the increases observed were substantially lower than predicted increases which assumed even mixing of sludge into 20 cm of top soil. For instance, at the highest rate an increase in the Cu concentration in soil of 222 mg/kg was predicted compared with the observed increase of 125 mg/kg.

Yields of herbage increased substantially as a result of sludge treatment. Concentrations of Cu in herbage from the trial are shown in Figure 3. The concentrations for each year are the average of two cuts. The greatest increase in Cu concentration occurred in 1977; from 4.6 mg/kg DM on the unsludged control plots to about 7 mg/kg in herbage from plots which had received as much as 443 kg Cu/ha added in sludge. In 1978, the highest concentration of Cu was found in herbage from the control plots (probably due to the presence of clover) and in 1979

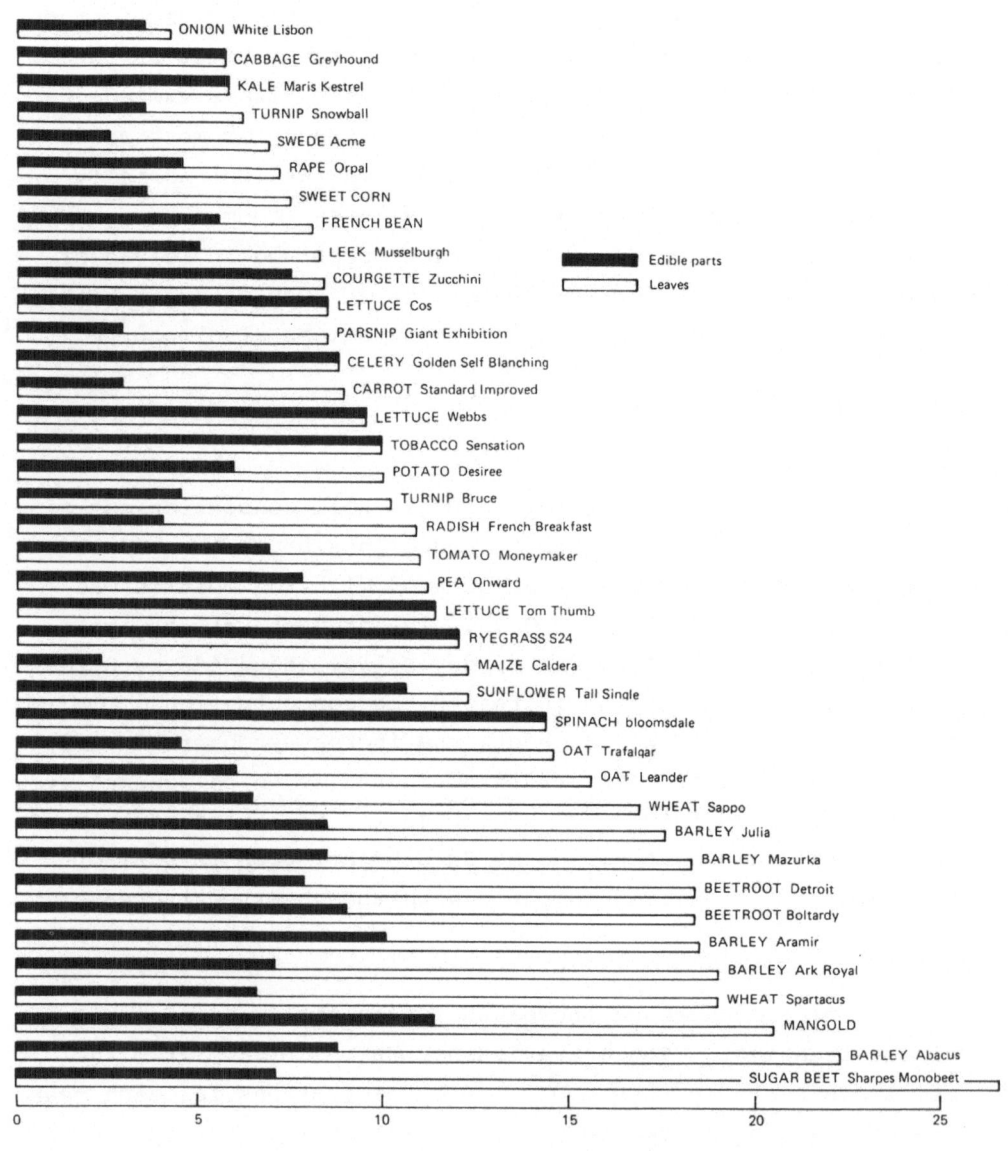

Fig. 1. Cu concentrations in crops grown on a single contaminated soil.

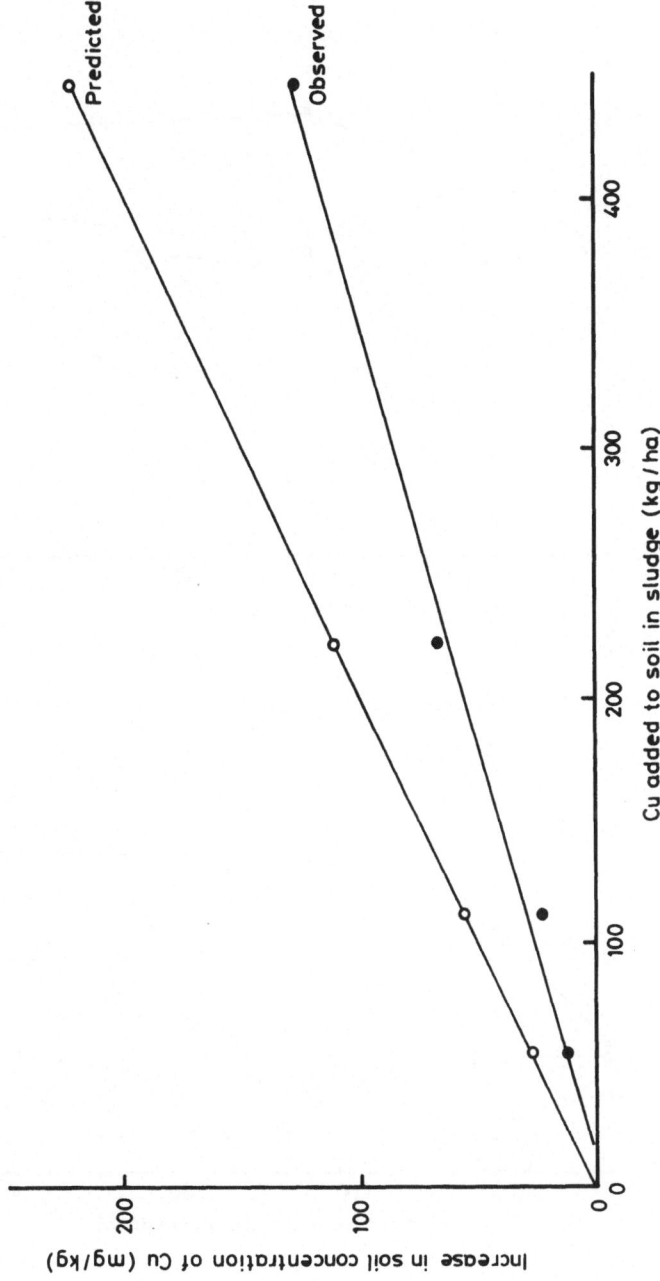

Fig. 2. Increases in soil concentrations of Cu resulting from Cu added in sludge.

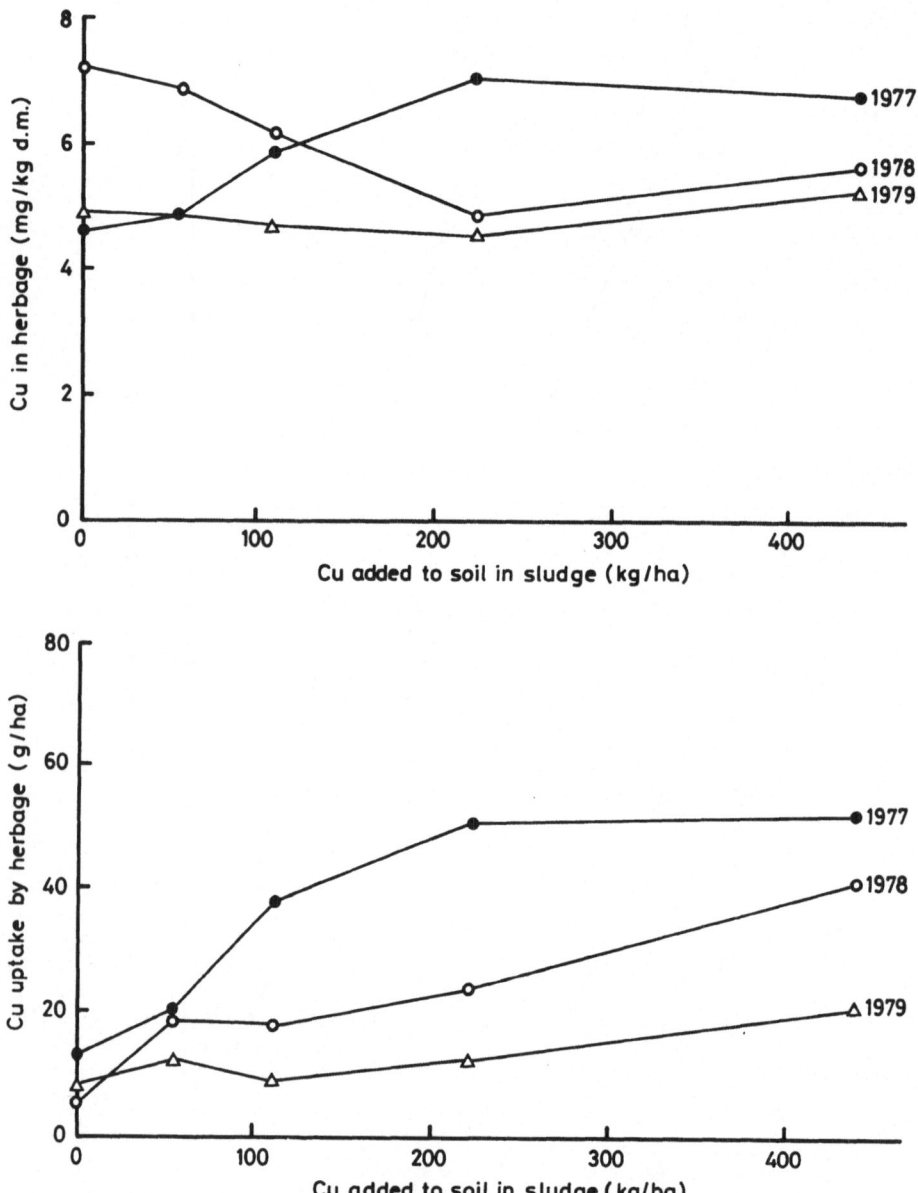

Fig. 3.   The relation between Cu added to soil in sludge and Cu in herbage.

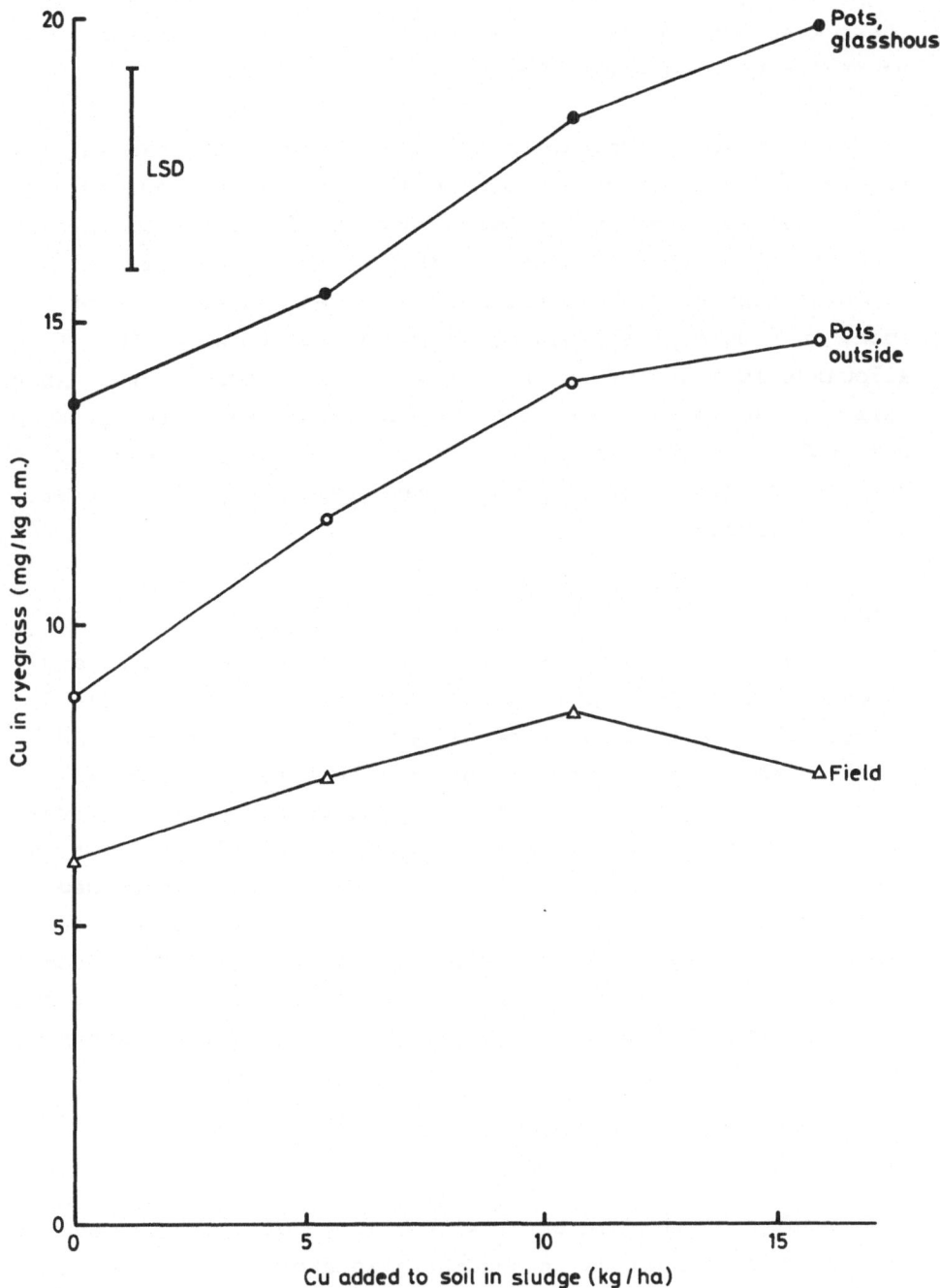

Fig. 4. Cu concentrations in ryegrass grown on sludge treated soil in different conditions.

there was little difference in the Cu content of herbage according to sludge treatment.

Results of Cu uptake in this trial are also shown in Figure 3. Greatest uptake occurred in 1977 when about 50 g/ha of Cu was recovered in the standing crop of herbage at the two highest rates of application of sludge which added 222 and 443 kg Cu/ha respectively to the soil. Percentage recovery of Cu added in sludge at the highest rate was 0.026, or 0.020% if allowance is made for uptake of Cu on the control plots. About half this amount was recovered in the first year after sludge was applied to the plots and thereafter uptake declined. Recovery of Cu in the crop was always less than 0.1% of that added in sludge.

Trial 3

Figure 4 shows that substantially higher concentrations of Cu occurred in ryegrass grown in pots compared with field plots. The least significant difference between means (LSD) shown on the graph was based on a probability (P) of 0.05. There was no significant increase in Cu concentration in ryegrass according to sludge treatment on the field plots but this was in contrast to pots. Here, Cu concentrations increased significantly with sludge treatment, the highest concentrations occurring in ryegrass grown in pots in the glasshouse. Even on unsludged soil, higher concentrations of Cu occurred in ryegrass grown in pots (8.8 mg/kg DM outside, 13.7 mg/kg glasshouse) than on field plots (6.1 mg/kg).

DISCUSSION

Results of the pot trial with a range of crops showed that considerable variations in concentrations of Cu may be expected to occur amongst different crops all grown on the same soil so that plant genotype is an important consideration in any assessment of the availability of Cu added to soil in sludge (Figure 1). Beets appear to be amongst the crops most sensitive to

soil Cu.   There is general agreement in the literature that
reductions in yield due to phytotoxicity may occur when Cu
concentrations in the leaves of crops exceed 20 mg/kg DM.   This
concentration was exceeded in the leaves of sugar beet, mangold
and barley (cv. Abacus) but the trial used a highly contaminated
soil in conditions likely to promote maximum uptake of Cu.

    Soil analyses from Trial 2 (Figure 2) suggest that
although additions of sludge increase soil concentrations of
Cu a simple relationship between the quantity of Cu added in
sludge and the concentration resulting in soil cannot be
assumed. Less than 0.1% of the Cu added to soil in sludge was
recovered in cut herbage.   Cu in digested sludge, in which it
occurs principally as sulphides and carbonates (Stover et al.,
1976), appears to be largely unavailable for plant uptake.
The greatest response was seen in the first year after sludge
was applied to the plots when an application of 443 kg Cu/ha
increased the Cu concentration in herbage by 2.2 mg/kg to 6.8
mg/kg DM.   This concentration is below the phytotoxic threshold
of 20 mg/kg and also below the level of 10 mg Cu/kg required in
the diet of cattle (ARC, 1965).   Uptake of Cu declined with
time after application of sludge to soil.

    Rates of application of Cu in sludge were lower in Trial 3
than Trial 2 (Table 1) but were made to a soil of lower pH
value where greater uptake of Cu might be expected to occur.
In field conditions there was no significant increase in Cu
concentrations in ryegrass in relation to sludge treatment but
with pots substantially higher concentrations of Cu in ryegrass
were found (Figure 4).   It seems unlikely that this marked
effect can be explained entirely by the split dressing of
sludge made to the field plots.   This finding confirms that of
De Vries and Tiller (1978) and indicates that the results of
pot trials should be interpreted cautiously.

    Average annual applications of sludge in the UK add
approximately 1.2 kg Cu/ha to grassland and 2.4 kg Cu/ha to
arable land.   The investigations reported above suggest that

application rates of this order will have no deleterious
effects on crop yields or the health of farm animals. Herbage
frequently contains less Cu than is required by ruminants (ARC,
1965) so that small increases in herbage content of Cu resulting from applications of sludge to grassland may be beneficial.
The question of direct ingestion of sludge and soil by grazing
animals is outside the scope of this paper. Nevertheless, it
is reassuring that Gracey et al. (1976) were able to conclude
that there would be little or no danger to the health of sheep,
which are sensitive to Cu toxicity, from grazing land which is
receiving pig slurry at rates which supply up to 16 kg Cu/ha/
year.

REFERENCES

ARC, 1965. Nutrient requirements of farm livestock, No. 2, Ruminants.
    Agricultural Research Council, U.K.

Berrow, M.L and Webber, J., 1972. Trace elements in sewage sludges. J.
    Sci. Fd Agric., 23, 93-100.

Davis, R.D. and Carlton-Smith, C.H., 1980a. The preparation of sludges of
    controlled metal content by the pilot-scale treatment of metal-
    enriched sewage, LR 1173. Water Research Centre, Stevenage.

Davis, R.D. and Carlton-Smith, C.H., 1980b. Crops as indicators of the
    significance of contamination of soil by heavy metals, TR 140.
    Water Research Centre, Stevenage.

De Vries, M.P. and Tiller, K.G., 1978. Sewage sludge as a soil amendment.
    Environ. Pollut., 16, 231-240.

Gracey, H.I., Stewart, T.A. and Woodside, J.D., 1976. The effect of
    disposing high rates of copper-rich pig slurry on grassland on the
    health of grazing sheep.   J. agric. Sci., Camb., 87, 617-623.

Lester, J.N., Harrison, R.M. and Perry, R., 1979. The balance of heavy
    metals through a sewage treatment works. Sci. Total Environ.,
    12, 13-23.

Matthews, P.J., 1980. Sewage sludge utilisation. Envir. Technol. Lett.,
    1, 65-80.

Stover, R.C., Sommers, L.E. and Silviera, D.J., 1976. Evaluation of metals
    in wastewater sludge. J. Wat. Pollut. Control Fed., 48, 2165-2175.

Thompson, L.H., 1979. Trade effluent control in London. In: Workshop on
    treatment of domestic and industrial wastewaters in large plants.
    IAWPR Conference, Vienna.

DISCUSSION

Th.M. Lexmond   *(Netherlands)*

I have two comments to make. The first one refers to
your Figure 1. As I showed this morning, oats are consistently
less sensitive to copper than, for example, spinach and
tomatoes, but they are very close together in your Figure 1.
Consequently, I do not think that the copper content of the
above ground part of the plant is indicative of the risk of
toxicity.

My second comment is with regard to Figure 4. It has
been shown repeatedly that the organic nitrogen content of the
shoots is of importance with respect to copper content. I can
imagine that differences in photosynthetic conditions and
nitrogen availability may have influenced the results presented
in Figure 4. Have you measured organic nitrogen, or protein
contents of the shoots of the ryegrass?

R.D. Davis   *(UK)*

On your first point: where sludge is used on the land,
the usual thing that is monitored is the concentration of metal
in the sludge, and possibly in the soil. I am regarded as
being mildly eccentric in some circles for suggesting that
people should look at the concentration of copper in the plant
tops. Were I to suggest that they should look at the concentration of copper in the plant roots I don't think I would
receive a very enthusiastic reception. Practical problems
associated with cleaning of roots make it impossible to obtain
reliable estimates of the copper content of plant roots grown
in field conditions (with the exception of swollen storage
roots). Although soil contamination may result in greater
increases in the copper content of roots compared with shoots,
any significant increase in the copper content of roots will
be matched by a smaller but measurable increase in the copper
content of the leaves. Unlike the copper in the roots,

copper in the shoots is likely to be physiologically active
and is of direct relevance to the nutrition of grazing
animals. The difference between your results for oats and
my results might be attributable to the different cultivars
used.

Th.M. Lexmond

But I think you are avoiding answering my question. You
suggested that there is a relationship between the copper
concentration in the above ground plant portions and sensitivity towards copper toxicity. I think that relationship
remains to be established. That was my first question which I
don't think you have answered.

R.D. Davis

I apologise for that omission. There is substantial
evidence in the literature to suggest that the copper concentration of leaves is a sensitive indicator of the copper status
of most plants. Dr. Beckett and I have published the results
of experiments investigating the effect of copper on crop yield.
In those experiments a concentration of 20 mg Cu/kg (dry matter)
in the leaves of crops was consistently associated with phytotoxic threshold as regards the onset of reduced yield due to
copper toxicity. Other investigations have found similar
results and copper concentrations in leaves are widely used for
diagnostic purposes, a concentration in excess of 20 mg Cu/kg
being taken to indicate possible toxicity.

In answer to your second point on Figure 4, we are
currently examining in detail the difference between growing
plants in the glasshouse and growing them outside on field
plots. I am interested in your suggestions concerning photosynthetic conditions and nitrogen availability and we shall
examine these factors.

T.W.G. Hucker (UK)

Figure 4 worries me a little bit because I find the very

great discrepancy between pots and field rather disturbing. Most of us knew there was quite a discrepancy but here we have discrepancies of 100%, which is somewhat alarming. May I ask Dr. Davis, is the Figure which refers specifically to copper typical of other metals, or is it just peculiar to copper?

R.D. Davis

Dr. Tiller has published some information in Australia which suggests that other elements are also affected. It seems likely that higher concentrations of various elements will occur in plants grown in glasshouse conditions compared with those grown in the field. At the same time, I think the effect is probably more marked for copper than for any other element.

J.H. Voorburg *(Netherlands)*

Is it significant that in the pot experiments all the roots are in contact with contaminated soil whereas in the field experiments the copper is only present in the top layer?

R.D. Davis

This comparison is slightly complicated because the field application of sludge involved two applications, one was ploughed in and one was applied to the surface. However, I would accept that that may be a factor although it would not explain the full effect.

D. McGrath *(Ireland)*

I would like to urge that more attention be paid to the nitrogen situation. This is very important for pot experiments or for once-off field experiments. It has been shown in the literature that the level of nitrogen actually in the soil, as distinct from the plant, affects the copper content of the herbage. I am thinking particularly of grass. In our experience, in pot experiments you may have copper contents elevated by as much as 5 ppm, purely by adding nitrogen. This is in controls. I would suggest that in Figure 4 it is strictly a matter of nitrogen availability. With sludge, of course, you add enormous quantities of nitrogen. This could also explain the effect between years. You had a massive dose of nitrogen in the first year. In the second year the available portion of that nitrogen would have been mineralised so that you then had less of a nitrogen interaction. The same applies to the third year.

R.D. Davis

That still leaves us with the problem of the control soil which had received no sludge. Higher concentrations of copper occurred in crops grown in glasshouse conditions for this soil too, compared with field conditions.

D. McGrath

You will get mineralisation in the glasshouse, even in your control soil.

R. Braude *(UK)*

What about the comparison between pots inside and pots outside? Your argument wouldn't apply there.

J.K.R. Gasser *(UK)*

Oh yes, there is a big change in temperature between inside and outside a glasshouse.

R. Braude

But not as far as nitrogen is concerned?

J.K.R. Gasser

Oh yes, very much so.

D. McGrath

And the whole watering regime is going to be totally different, the transport of nitrogen to roots and so on, it is all going to affect the amount of nitrogen available to the plant.

Th.M. Lexmond

Following Dr. McGrath's comment, Dr. Beyme in Germany advised us as long ago as 1971 to base the copper content of a crop on its organic nitrogen content, instead of on dry matter.

R.D. Davis

I am interested to hear that; it does not seem to be an idea that has caught on since 1971. Can you give me the reference to this work?

J.K.R. Gasser

It is not only a question of absolute amount of nitrogen, it is the form of nitrogen as well. Nitrogen mineralises first of all into the ammonium form. If it remains there for any length of time it will be nitrified. One of the things that has fascinated me for years, and I don't know the answer, is what is the effect of the form of nitrogen, ammonium or nitrate, on copper uptake by plants? My theory tells me that ammonium ought to have a synergistic effect. I would love to see it proved.

Th.M. Lexmond

I will follow your suggestion and take a look at that point.

A. Dam Kofoed   (Denmark)

I am pleased to see the comparison you show in Figure 4 of results obtained in pot trials and in field experiments. The difference is probably partly due to the greater density of plant roots in a given volume of soil which would be found for plants grown in pot culture compared with field conditions. Your results confirm the questionable relationship between results obtained from pot trials compared with field trials.

R.D. Davis

I am not suggesting that pot trials should be stopped but

merely that we should be careful in the way that we use and interpret them; they are very useful for some purposes.

G.A. Fleming (Ireland)

I think it is almost impossible, and indeed foolish, to endeavour to compare pot trials and field trials. I would go so far as to say that. Pot trials serve to show a particular thing; you may get a completely opposite, though correct, result in the field. I once did a pot trial on selenium and its effect as the plant went to maturity. In the pot, the selenium content rose as the plant matured. When the same experiment was done in the field, the selenium content declined. Two totally different results. Of course plot experiments should not be knocked but to try to compare a pot experiment to a field experiment at any time is foolish.

K.L. Robinson (UK)

I would just like to draw attention to the fact that in the many experiments that have been published in which the effects of copper enriched pig slurry on the land are compared, they are not true comparisons. What is compared is a control with chemical fertiliser with plots containing slurry with copper in it. I have seen figures for experiments with a better comparison, that is to say, controls with chemical fertiliser, controls with slurry with no added copper from pig feeding and the usual copper enriched pig slurries. You do find a difference. You find the slurry plots with no copper are intermediate between the two. The same thing applies to sludges. Nobody is comparing sewage sludge with no copper in it with sewage sludge that has copper. What you are comparing is sewage sludge with copper with chemical fertiliser controls. It is not really a valid comparison.

R.D. Davis

That is a very fair point; it is something we are beginning to tackle.

J.B. Ludvigsen *(Denmark)*

I notice from your Figure 1 that beetroots are one of the crops most sensitive to copper toxicity. Is this because beetroots store most of their nitrogen as nitrate in the root?

R.D. Davis

This might be a factor because we have heard in discussion that the nitrogen status of soils and plants is intimately connected with copper metabolism. The reasons for the varying ability of different plants to assimilate copper deserve further investigation.

M.D. Webber *(Canada)*

Do you have any suggestion as to what is happening to your copper in the soil? Is it going down the profile?

R.D. Davis

Well, we are losing it as is shown in Figure 2. But this was a heavy soil with a high pH value, so I find it hard to believe that any significant amount would be leached. I think perhaps the activity of earthworms may be quite important here. Worm populations build up following additions of organic matter to soil and Charles Darwin showed in the last century that worms turnover large quantities of soil during their burrowing activities. This may physically move copper down the profile and is a mechanism to consider in addition to leaching.

J.K.R. Gasser

There is another very interesting facet to that which is that the roots tend to accumulate copper and the worms feed on the roots. So they tend to get far more copper than one might think in general terms.

SESSION II

AGRONOMY

B) PLANT/ANIMAL RELATIONSHIPS

Chairman: K.L. Robinson

EFFECTS OF THE DISPOSAL OF COPPER-RICH SLURRY
ON THE HEALTH OF GRAZING ANIMALS

I. Bremner
Nutritional Biochemistry Department, Rowett Research Institute,
Bucksburn, Aberdeen, UK.

ABSTRACT

   The effects of the disposal of pig slurry of high copper content on the development of copper toxicosis in sheep are reviewed. Special consideration is given to the increases in pasture copper content which can arise from surface contamination of plants with slurry residues. Studies on the absorption and hepatic accumulation of slurry-bound copper by sheep indicate that this copper is relatively available to animals. However, no cases of copper toxicosis or any major increases in liver copper content have been reported when sheep were allowed to graze severely contaminated pasture for up to 3 years. Provided reasonable precautions are taken as to the rates and times of slurry application, there appears to be no immediate hazard to grazing sheep from slurry disposal.

INTRODUCTION

The demonstration of the growth-promoting effects of copper in pigs has led in many countries to the routine addition of 200 - 250 mg copper/kg of pig diets. Provided the dietary content of other metals such as zinc and iron is adequate, then these amounts of copper are not toxic to the pigs, even though they are far in excess of the pig's minimum dietary requirement for that metal. Only a small proportion (about 5%) of the copper in these rations is retained by the pigs and most of the metal is excreted in the faeces, with the result that the slurry from pig-rearing units generally contains about 700 mg copper/kg dry matter. According to some environmentalists, repeated disposal of this slurry on farmland may eventually have deleterious effects on soil composition and on plant production. It may also constitute a hazard to animals grazing slurry-treated pastures or receiving conserved herbage from these pastures, as is indicated by the report of the death from copper poisoning of sheep receiving hay from contaminated pastures (Feenstra and Ulsen, 1972).

There is some uncertainty, however, as to the magnitude of this latter hazard, despite the publication of several reports on the topic. It is the aim of this paper to review these reports and to assess whether the spreading of copper-rich pig slurry on farmland really does represent a significant hazard to animals.

COPPER TOXICITY IN SHEEP

Sheep are more likely to be affected by the disposal of copper-rich slurry than are cattle, since the former species is much more susceptible to chronic copper poisoning (for reviews see Todd, 1969; Howell, 1977; Bremner, 1979). The development of chronic copper poisoning is characterised by two distinct phases. In the first of these there is a gradual accumulation of copper in the liver over a period of several months. During this time there are no overt signs of disease

and growth rates and food intakes are generally normal. There
may however be sub-clinical liver damage, as can be demonstrated
by histological and electronmicroscopical examination of the
liver and by measurement of the activities in the plasma of
liver-specific enzymes. The onset of the second and terminal
phase of the disease occurs relatively suddenly and is
characterised by a large increase in blood copper concentractions
and by extensive haemolysis of the red blood cells. This
'haemolytic crisis' is accompanied by methaemoglobinaemia and
haemoglobinuria, and usually has fatal consequences.

In one experiment in which growing lambs received a
barley-fishmeal diet with 29 mg Cu/kg, 40% of the animals
developed the haemolytic crisis after 17 - 22 weeks (Bremner
et al., 1976). Liver copper concentrations at the end of the
experiment were about 400 mg/kg fresh weight. However, liver
damage was evident after only 12 weeks (Bremner et al., 1976;
King and Bremner, 1979), the severity being related to the
liver copper concentrations, which ranged at this time from
120 to 440 mg/kg fresh weight (mean 240 mg/kg).

Unfortunately it is not possible to define critical liver
or dietary copper concentrations at which the haemolytic crisis
and other signs of copper poisoning will invariably occur,
although the above figures are fairly typical of other findings
(e.g. Tait et al, 1971; MacPherson and Hemingway, 1969). Onset
of the haemolytic crisis is not determined solely by liver and
copper concentration and other factors, including stress, are
also involved. Although the dietary copper intake is un-
doubtedly a major determinant of liver copper accumulation in
sheep, there are important effects of breed, age and dietary
composition on the absorption and retention of copper and on
the development of copper toxicosis (see Bremner, 1979). For
example, the onset of the haemolytic crisis and the hepatic
accumulation of copper have been effectively controlled in
sheep, even at high dietary copper intakes, simply by
increasing the zinc or molybdenum content of the ration
(Bremner et al., 1976; Suttle, 1978).

The existence of such diverse factors which can influence the development of copper toxicosis obviously makes it extremely difficult to assess definitively the hazard which is posed by the disposal of copper-rich slurry on pasture. However, it will be assumed in the first instance that a problem is only likely to occur if the copper intake is 25 - 30 mg/kg diet and/or the liver copper content is about 300 mg/kg fresh weight.

COPPER CONTENT OF SLURRY-TREATED PASTURES

It seems unlikely that the disposal of slurry will so enhance the uptake of copper by plants that such dietary concentrations will readily be attained, provided reasonable levels and systems of application are used. In one experiment by Price (1979), where care was taken to minimise the degree of surface contamination of ryegrass plots by cutting the grass immediately prior to each application of slurry, mean copper concentrations in the plant increased by only 3 - 4 mg/kg dry matter. The unprocessed slurry, which contained about 700 mg copper/kg dry matter, was applied at the relatively high rate of 50 000 litres/ha on each of 7 occasions over a 2 year period. The total amount of copper applied was 26 kg/ha. The change in ryegrass copper concentration was variable and depended on the nature of the soil, the stage of development of the plant and the degree of dilution of the slurry. However, the mean copper concentration in the ryegrass increased from 4.57 ± 0.3 to only 8.15 ± 0.57 mg/kg on one soil and from 7.47 ± 0.33 to 11.5 ± 0.93 mg/kg on another soil.

Batey et al. (1972) reported somewhat greater increases in pasture copper levels in a comparable small-scale experiment where pig slurry was applied to two grass plots at rates of 56 000 and 112 000 litres/ha on 3 separate occasions during the growing season. This supplied a total of 6.1 and 12.2 kg copper/ha. Under these conditions mean plant copper concentrations throughout the year were 9.1 mg/kg in a control plot and 18 and 21 mg/kg respectively in the sprayed plots.

Individual values varied greatly and were on occasion as high as 40 mg/kg. However, these only occurred when the grass was sampled soon after application of the slurry, implying that there was appreciable surface contamination of the herbage. This view was supported by the finding that copper concentrations could be reduced by about half on washing of the sample before analysis.

A survey of the copper levels in pastures from commercial or experimental farms where slurry was applied regularly over a three year period (Table 1) confirms that considerable variations can occur in copper concentration, since these ranged from 8 to 164 mg/kg. The highest values were again only obtained when samples were collected soon after slurry application. Concentrations generally decreased after a period of heavy rainfall and were relatively high (about 30 mg/kg) when the cut grass was collected by forage harvester when the weather was dry, presumably because of uptake of slurry residues from the soil surface (Dalgarno and Mills, 1975).

TABLE 1

HERBAGE COPPER CONCENTRATIONS AFTER APPLICATION OF PIG SLURRY TO PASTURES FOR 3 YEARS

Level of application		Herbage copper	Reference
litres/ha/yr	kg Cu/ha/yr	(mg/kg)	
730 000	12.5	9.6 - 16.1	Batey et al., 1972
66 000	-	8 - 13	Dalgarno and Mills, 1975
315 000	16	164 (1d postapplication)	Gracey et al., 1976
		84 (14d " ")	
		50 (28d " ")	

AVAILABILITY OF COPPER FROM SLURRY TO SHEEP

Some of these copper concentrations in the grass samples contaminated with slurry are sufficiently high to come within the range where problems might be expected to arise in grazing

livestock. However it cannot be assumed that the copper
deposited on the leaf surface will necessarily be as available
to animals as copper incorporated into plant tissue. Indeed
if the deposited copper were present as CuS, as was once
suggested, then very low availability would be expected. More
recent studies suggest that the copper in pig slurry is bound
principally to organic matter (Price, 1979) and is, in fact,
highly available to sheep (Dalgarno and Mills, 1975; Suttle
and Price, 1976).

The availability of copper was measured by monitoring
the change in plasma copper and caeruloplasmin concentrations
when copper-deficient ewes or lambs were repleted with a semi-
purified diet containing 1 - 2% of dried pig slurry (Suttle
and Price, 1976) or the solid residue from an aerobic digest
of slurry (Dalgarno and Mills, 1975). No significant
differences were noted in plasma response between animals
receiving copper in the form of slurry or as $CuSO_4$. However
hepatic copper retention in the slurry-fed lambs was only about
half that in those receiving $CuSO_4$ indicating some decrease in
copper availability from the slurry (Dalgarno and Mills, 1975).
Nevertheless liver copper concentrations in the slurry-fed
animals were as high as 267 ± 19 and 474 ± 24 mg/kg dry matter
(approximately 75 and 130 mg/kg fresh weight) after the diets
containing 1 and 2% slurry residues (13.4 and 23.6 mg copper/kg
diet) had been fed for 4 months. Clearly prolonged ingestion
of diets contaminated with high levels of slurry could eventually increase liver copper concentrations to levels where
copper toxicosis might develop. Mills and Dalgarno (1975)
suggested, for example, that it would take about 8 months for
liver copper to increase to 1 000 mg/kg dry matter, if slurry
residues accounted for 2% of the sheep's dry matter intake.

Fortunately it is unlikely that ingestion of such large
amounts of slurry residues would occur continuously in grazing
animals for such a prolonged period, since the degree of contamination of the pasture decreases appreciably with time after
slurry application, especially after heavy rainfull. In these

circumstances, the most reliable way to assess the risk of copper poisoning is probably to monitor directly the changes in liver copper concentrations and the onset of liver damage in animals grazing severely contaminated pasture for a prolonged period.

Gracey et al (1976) determined the effects on grazing sheep of the repeated disposal of pig slurry over a 3-year period onto a permanent grassland sward. A total of 900×10^3 litres of slurry were applied/ha at monthly intervals during the growing seasons (45 000 litres/ha/application), equivalent to a total of 47 kg copper/ha. The pasture became severely contaminated with slurry, with mean herbage copper concentrations decreasing from 150 mg/kg dry matter to 50 mg/kg during the period between slurry applications. Despite the fact that the same animals were allowed to graze these pastures over the 3-year period and were also subjected to the stress of pregnancy, no deaths from copper poisoning were reported in the ewes or their lambs. Liver copper concentrations in the ewes at slaughter were only 150 mg/kg wet weight, compared with values of 134 in animals grazing control pasture which was not treated with slurry. Corresponding values in the lambs were 34 and 85 mg/kg.

Considerable increases in serum GOT levels were noted during the second grazing season, which could indicate that the increased copper intake had induced liver damage. However this increase was transient, only occurred when the grass supply was limited, and was also observed to some extent in the control animals. It is likely, therefore, that factors other than copper ingestion were responsible for the increased GOT levels.

Kneale and Smith (1977) also reported increased levels of GOT, sorbitol dehydrogenase and of glutamate dehydrogenase in the blood of ewes grazing slurry-treated pasture. Slurry was applied at a high rate (250 000 l/ha) during the winter and on several occasions during the grazing season to give pasture copper concentrations of about 175 mg/kg in June/July and

30-40 mg/kg in August/September. Although the increased blood enzyme levels could indicate the occurrence of copper-induced liver damage, this is difficult to reconcile with the absence of a major increase in liver copper in the lambs. Concentrations in the animals grazing slurry-treated and control pastures were only 120 and 25 mg/kg dry matter respectively. This is far below the level at which serious liver damage would be expected.

The absence of any major increase in liver copper content in these grazing trials is surprising in view of the relatively high availability of the copper in slurry residues when fed directly to sheep. Dalgarno and Mills (1975) have suggested however that this could be a reflection of differences in the conditions to which slurry residues were exposed between voiding by pigs and ingestion by sheep.

CONCLUSION

None of the above investigations into the effects of the disposal of slurry on farmland has provided convincing evidence that a serious problem exists at present to grazing sheep. This view is supported by the fact that there have been few reports of copper toxicosis developing on commercial farms as a direct consequence of the disposal of copper-rich slurries, even though high-copper pig diets have been routinely used throughout the world for over two decades. This does not mean that indiscriminate disposal of the slurry can be sanctioned but it does suggest that, with proper attention to pasture management, the hazard is at present small. Provided care is taken to minimise the degree of surface contamination of the plant with slurry, by applying the slurry when the grass is short and at a reasonable length of time before animals are allowed to graze, then pasture Cu levels should remain within a reasonable and acceptable range. If grass is to be conserved however, then it should be remembered that the use of a forage harvester can, under dry conditions, result in appreciable uptake of slurry residues from the soil surface.

Unfortunately, it is not known whether the continued disposal of these slurries will ultimately increase the levels of available copper in the soil to such an extent that plants can no longer limit the uptake of copper into the foliage. It is known that the continued disposal for 70 years of high-copper residues from whisky distilleries has caused some increase in plant copper concentrations in some areas of North-East Scotland (Reith et al., 1979). For example when ryegrass was grown on contaminated soil containing about 300 mg of EDTA-extractable copper/kg, the copper content of the grass increased to about 17 mg/kg. Concentrations in clover were about 30% greater than this, and were thus near the range where its exclusive use as a feedstuff for sheep could be dangerous. It has been claimed however that copper is phytotoxic to clover at soil concentrations of available copper of only 30 mg/kg (Purves, 1977), which suggests that effects of continued slurry disposal on plant growth and perhaps on soil composition will be of much greater importance than possible effects on grazing animals.

REFERENCES

Batey, T., Berryman, C. and Line, C., 1972. The disposal of copper-
    enriched pig-manure slurry on grassland. J. Brit. Grassld Soc.
    27, 139-143.

Bremner, I., 1979. Copper toxicity studies using domestic and laboratory
    animals. In: Copper in the Environment Part II. (J.O. Nriagu, ed.)
    pp. 286-306. (Wiley Intersciences, New York).

Bremner, I., Young, B.W. and Mills, C.F., 1976. Protective effect of zinc
    supplementation against copper toxicosis in sheep. Brit. J. Nutr.
    36, 551-561.

Dalgarno, A.C. and Mills, C.F., 1975. Retention by sheep of copper by
    aerobic digests of pig faecal slurry. J. Agric. Sci. 85, 11-18.

Feenstra, P. and Ulsen, F.W. Van, 1973. Hay as a cause of copper poisoning
    in sheep. Tijdschrift voor Diergeneeskunde 98, 632-633.

Gracey, H.I., Stewart, T.A., Woodside, J.D. and Thompson, R.H., 1976. The
    effect of disposing high rates of copper-rich pig slurry on grass-
    land on the health of grazing sheep. J. Agric. Sci. 87, 617-623.

Howell, J. McC., 1977. Chronic copper toxicity in sheep. In: The
    Veterinary Annual, 17th Issue (C.S.G. Grunsell & F.W.G. Hill, eds.)
    pp. 70-73 (Wright Scientechnica, Bristol).

King, T. and Bremner, I., 1979. Autophagy and apoptosis in liver during
    the prehaemolytic phase of chronic copper poisoning in sheep.
    J. Comp. Pathol. 89, 515-530.

Kneale, W.A. and Smith, P., 1977. The effects of applying pig slurry
    containing high levels of copper to sheep pastures. Exp. Husb.
    32, 1-7.

MacPherson, A. and Hemingway, R.G., 1969. The relative merit of various
    blood analyses and liver function tests in giving early diagnosis
    of chronic copper poisoning in sheep. Brit. Vet. J. 125, 213-221.

Price, J., 1979. The effects and fate of copper from pig slurry when
    applied to soil. Ph.D. Thesis, University of Edinburgh.

Purves, D., 1977. Availability of trace elements in soil. In Trace
    element contamination of the environment: Fundamental aspects of
    pollution control and environmental science. Series 1. pp 121-148
    (Elsevier North Holland Inc., New York).

Reith, J.W.S., Berrow, M.L. and Burridge, J.C., 1979. Effects of copper in distillery wastes on soils and plants. In Management and Control of Heavy Metals in the Environment. pp. 537-540. (CEP Consultants, Edinburgh).

Suttle, N.F., 1977. Reducing the potential copper toxicity of concentrates by the use of molybdenum and sulphur supplements. Anim. Feed Sci. Technol. 2, 235-246.

Suttle, N.F. and Price, J., 1976. The potential toxicity of copper-rich animal excreta to sheep. Anim. Prod. 23, 233-241.

Tait, R.M., Krishnamurti, C.R., Gilchrist, E.W. and MacDonald, K., 1971. Chronic copper poisoning in feeder lambs. Canad. Vet. J. 12, 73-75.

Todd, J.R., 1969. Chronic copper toxicity of ruminants. Proc. Nutr. Soc. 28, 189-198.

DISCUSSION

M. Lamand *(France)*

In relation to your last slide, what was the level of copper in the field contaminated by slurry and copper sulphate?

I. Bremner *(UK)*

I can't remember the values quoted for the copper sulphate. For the slurry, it ranged from about 160 ppm immediately after the slurry application down to about 50 ppm one month afterwards. When the next batch of slurry was applied it went back to 160 ppm or thereabouts. The animals were grazing throughout this period; I think they were taken in during the winter.

J.B. Ludvigsen *(Denmark)*

In the slide where you showed the concentration of copper in the liver of the sheep on slurry, and there was an increase in liver of the lambs from 34 up to 80 ppm, could that be due to the fact that the sheep excrete copper through the milk, or were the lambs grazing?

I. Bremner

The lambs were grazing.

R. Braude *(UK)*

I would like to support your statement from Gracey's experiment. There are very interesting reports from America which have been published by Prince, Hays and Cromwell from Kentucky, on experiments where slurry from pigs fed 150 ppm copper, was continuously applied to the land for three years. Lambs were grazed on that land continuously throughout that period without any harmful effect. Eventually some of the sheep were slaughtered and the liver copper levels examined. These workers support the idea that there is an unknown factor which may precipitate toxicity but they couldn't establish what that factor was even over three years. So this is confirmation from a very large experiment which has been published.

J.H. Voorburg *(Netherlands)*

I have asked two of the animal health services in our country about the situation with copper toxicity in sheep. No exact figures were available but they said it was quite common. It seems that there is also a correlation between copper toxicity and the number of pigs on the farm, and apparently, most cases of copper toxicity occurred during the period in which the lambs are born when the sheep have extra concentrates. A possible hypothesis may be that the combination of grazing on grass which has been treated with copper enriched pig slurry and the feeding of concentrates which contain copper causes copper toxicity.

I. Bremner

This is certainly a possibility. The development of copper poisoning in sheep in the UK is only a problem in intensive breeding systems.   My view is that it is the con-centrates which are being fed which are responsible, since it is not uncommon for their copper concentrations to be around 25 to 30 ppm.   Moreover, as Professor Todd pointed out, associated with this high copper intake, you have a low moly-bdenum content in cereal grains. As molybdenum tends to inhibit copper utilisation a low molybdenum intake is there-fore a bad thing if you think you are in danger of copper poisoning.   There is probably a combination of these two effects.

B.C. Cooke *(UK)*

Would you suggest that zinc and molybdenum should be added to the compound feed in order to control potential copper poisoning?

I. Bremner

There is certainly no harm in increasing zinc up to about 200 ppm. With regard to molybdenum, we know that molybdenum levels of only a few ppm are sufficient to inhibit copper utilisation.   If you are feeding concentrates it is most

unlikely that you would ever encounter copper deficiency. So
I suppose one could sanction a level of addition of around 2
or 3 ppm for molybdenum.

K.L. Robinson (UK)

With regard to the point that Dr. Voorburg mentioned,
there was a report from the Dutch Animal Health Commission
about the relationship of pigs kept on farms to sheep poisoning. When you look at the data you find that the deaths were
in mid winter and the sheep were fed indoors, not grazed on
pasture. They were fed with forages harvested from the pastures which had been treated. I wonder if Dr. Bremner could
comment on that. There may be some physiological point involved
with indoor feeding of sheep with high copper rations compared
with outdoor grazing.

I. Bremner

This is possible but I think it is often a question
mainly of copper intake. As you know, in the UK copper levels
in pastures are generally quite low. We have heard already
that quite often they are insufficient to meet the demands of
copper by cattle. The switch from the situation where the
copper intake is only marginally adequate to that of feeding
concentrates with a high copper content, is probably the main
factor responsible for the development of copper poisoning in
sheep. However, I admit that we do not know which other factors
are involved in the onset of the haemolytic crisis.

R.J. Unwin (UK)

It has been shown, and it was mentioned this morning,
that there are dangers in incorporating slurry residues into
conserved forages. I think Dr. Robinson was talking particularly about conserved feeds in that instance; there can be a
tie-up but, again, I would consider it a management factor in
the way in which the forages are made.

I. Bremner

This is a point that was brought out in the paper by
Dalgarno and Mills. They noted that some animals at the Rowett
which had been receiving a dried grass ration showed very high
live copper concentrations. This was traced back to the source
of the dried grass which was a farm which had a large pig unit
and which disposed of a lot of slurry on the land. What had
apparently happened was that the slurry residues were being
picked up at forage harvesting and were increasing the copper
content of the grass at this point. I think the level of
copper in the dried grass was as high as 30 ppm.

D.B.R. Poole *(Ireland)*

This may be at slight variance with what Dr. Bremner has
said but I think that in concentrate fed sheep, the level of
copper that you are talking about is higher than is absolutely
necessary to produce copper toxicity. We have certainly seen
copper toxicity occurring in young sheep where they were fed
over an extended period of time and I don't believe the levels
were as high as 30 ppm. I would put them at between 10 and 15
ppm, maybe even lower.

Is there any information about the relative susceptibility of the Texel breed of sheep vis-a-vis other European
breeds? I ask this because recently we had two suspected out-
breaks of copper toxicity in Texel sheep in Ireland under a
situation where I would not have expected copper toxicity under
our own breeding conditions.

I. Bremner

As I said yesterday, there are marked breed differences
in sheep in terms of susceptibility to copper deficiency and
copper poisoning. This is shown by work done by Dr. Wiener at
the Animal Breeding Research Organisation in Edinburgh. I am
also aware of one report from Germany by Luke and Wiemann who
looked at some European breeds of sheep and found that the
Texel breed was the most susceptible to copper poisoning.

A.D. Bird *(UK)*

Would Dr. Bremner care to comment on the situation in relation to the level of added copper permitted in manufactured feedingstuffs in the UK, and in the rest of the Community? In the UK we prohibit added copper in manufactured feeds for sheep. In the Community regulation, up to 50 ppm copper is permitted in sheep rations. Is it a case of the UK being too cautious, or is it a case of the rest of Europe being too lax?

I. Bremner

Well, I will put my neck on the block and say that I think it is a case of the rest of Europe being too lax. I would suggest that to sanction the addition of 50 ppm copper to sheep rations is to ask for trouble. I cannot see the justification for it when the copper requirement is thought to be around 5 ppm.

K.L. Robinson

This issue has been raised in the EEC Animal Nutrition Committee that is considering the whole question of copper in livestock diets. In fact, my function here is as a liaison between that committee and this workshop. Without wishing to anticipate the conclusions and recommendations I do feel that 50 ppm addition for sheep diets may be the cause of some trouble.

COPPER TOXICITY IN SHEEP

M. Lamand

Laboratoire des Maladies Nutritionnelles
INRA - CRZV de Theix,
63310 Beaumont, France.

ABSTRACT

From field observations it is apparent that 15 mg of copper per kg of dry matter in French forages is the upper limit beyond which copper toxicity, mainly in lambs, is likely to occur. This relatively low value is attributed to low molybdenum concentrations in French feeds.

Copper toxicity in sheep is characterised by a haemolytic icterus crisis after the accumulation of large amounts of copper in the liver. The diagnosis of a copper toxicity is based mainly on liver copper concentrations (> 1 000 mg Cu/kg DM) and copper and molybdenum levels in the diet.

Prophylactic measures include the feeding of copper deficient forages alone, or in combination with sulphur and molybdenum supplementation. Supplementation of the diet with zinc may also be effective in the prevention and treatment of copper toxicity.

INTRODUCTION

The widespread use of copper supplements in the diets of pigs has created problems in the disposal of pig manure due to its high copper content. An attempt is made, with particular reference to copper and molybdenum concentrations of French forages, to evaluate the possible risk to sheep grazed on pastures sprayed with pig manure high in copper.

It is therefore important to monitor closely any appearances of clinical symptoms of a copper toxicity and to organise an appropriate means of prevention and treatment.

COPPER TOXICITY LIMITS IN SHEEP

Sheep differ from most other animal species in their low tolerance to dietary copper concentrations.

Copper deficiency, associated with swayback disease in sheep has been demonstrated in France. The preventive measures employed include the supplementation of the diet with copper, however, the determination of dietary limits of copper (minimum and maximum) is influenced by dietary molybdenum concentrations in view of the fact that molybdenum is a copper antagonist.

The accumulation of toxic levels of copper in the liver is achieved when there is an excess of copper in the diet. This phase may last for several weeks.

Molybdenum tends to limit the absorption of copper from the gut and to lower its accumulation in the liver and limit utilisation by cells. The quantity of sulphur in the diet is also an important factor in copper metabolism since sulphur has been found to be present in the di- and trithiomolybdates which are probably the compounds formed in the rumen and which are involved in a complex copper-sulphur-molybdenum interaction (Clarke and Laurie, 1980). These two compounds are absorbed from the intestinal tract of sheep and tend to lower caeruloplasmin activity (Mason et al., 1980).

Due to the complexity of the copper-sulphur-molybdenum interaction, copper toxicity in sheep may appear at rather low copper levels if dietary molybdenum concentrations are low (< 0.2 mg Mo/kg DM).

According to our observations of some naturally occurring cases of copper toxicity, the dietary limit for copper toxicity, mainly in lambs, is approximately 15 mg Cu/kg DM) (Table 1).

TABLE 1

FIELD OBSERVATIONS OF COPPER INTOXICATION IN SHEEP

Animals	Dietary trace elements (mg/kg DM)	Symptoms	Time before appearance
Weaned lambs	Cu : 14 - 18 Zn : 85 - 94 Mn : 78 - 98 Mo : 0.6 - 0.8	Death Rejected after slaughtering for icterus	70 - 80 days
Ewes	Cu : 40	30% death after haemolytic crisis and icterus	70 days
Ewes	Cu : 20 - 25	Chronic anaemia (haematocrit 20 - 30%)	One year

Most of the hay made from natural French pastures (1st cut) is copper deficient (Table 2) as well as low in molybdenum (Figure 1). The low molybdenum concentrations enhanced the risk of copper toxicity in sheep especially when pig manure rich in copper was sprayed onto pastures. Rye grass copper and zinc concentrations rose to 20 mg/kg DM and 150 mg/kg DM respectively when copper and zinc were added to soils at the rate of 500 kg/ha (Coppenet, 1980) which was the beginning of a phytotoxic state. The main risk therefore is the direct contamination of the forage by copper originating from the pig manure. It has been demonstrated that copper originating from

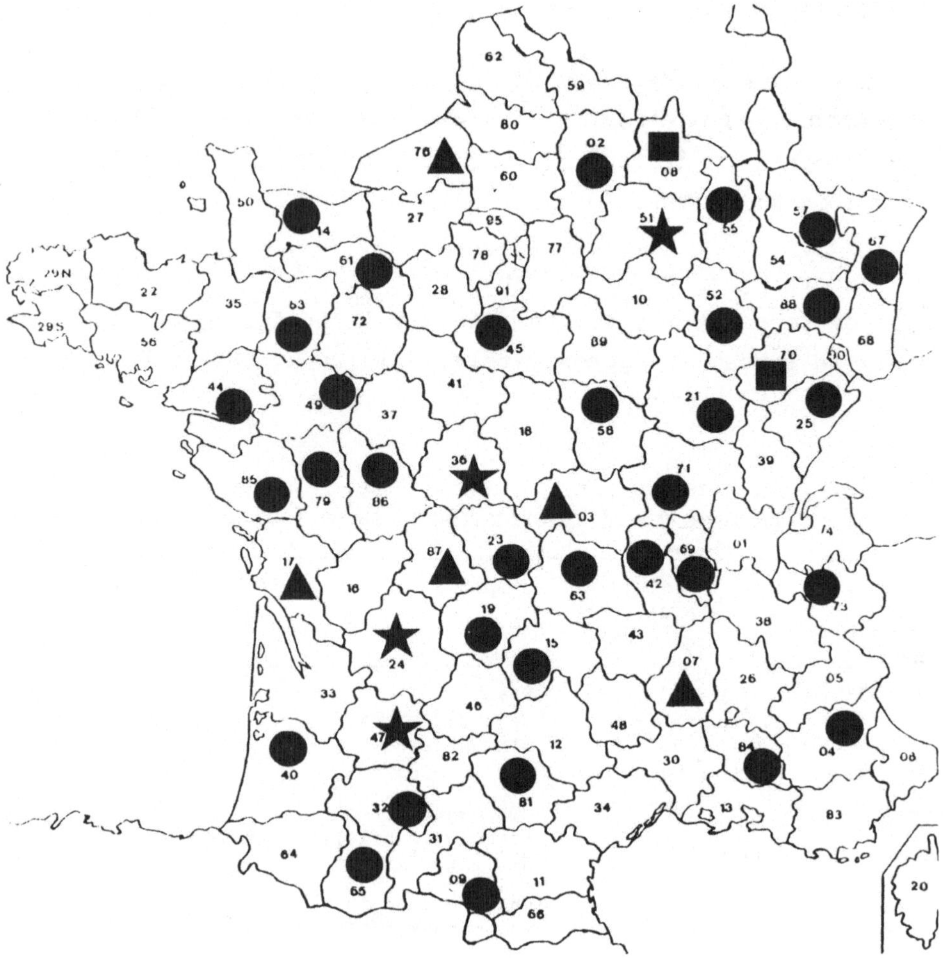

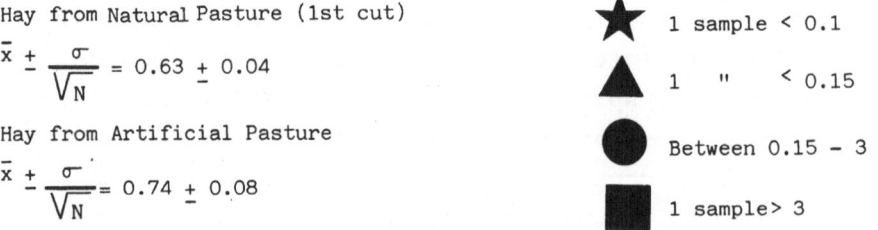

Hay from Natural Pasture (1st cut)

$$\bar{x} \pm \frac{\sigma}{\sqrt{N}} = 0.63 \pm 0.04$$

Hay from Artificial Pasture

$$\bar{x} \pm \frac{\sigma}{\sqrt{N}} = 0.74 \pm 0.08$$

★ 1 sample < 0.1

▲ 1 " < 0.15

● Between 0.15 - 3

■ 1 sample > 3

Fig. 1. Concentration and geographical distribution in France of
molybdenum in the hays (mg/kg DM)

the pig manure is as biologically available to sheep as that from copper sulphate (Suttle and Price, 1976).

TABLE 2

DISTRIBUTION OF COPPER IN NATURAL FRENCH PASTURE HAY (FIRST CUT)

Copper concentrations (mg/kg DM)	% total hays sampled
< 5	44.2
5 - 7	49.5
> 7	6.3

CLINICAL ASPECTS OF COPPER TOXICITY IN SHEEP

Symptoms

Copper toxicity evolves in two phases :

1. copper accumulates more or less slowly in the liver, until the appearance of a haemolytic icterus crisis

2. the second phase is characterised by a sudden increase in the concentration of copper in the blood and most animals die within a short period thereafter.

The liver copper accumulation phase (phase 1) is shorter with higher dietary intakes of copper and lower dietary intakes of sulphur and molybdenum. If the copper intakes are very close to the toxicity limit the accumulation phase may be very long with anaemia being the only sign of a copper toxicity.

The symptoms and necropsy findings of copper toxicity have recently been reviewed by Soli (1980) (Table 3). It was pointed out that most specific symptoms are haemolytic icterus and the emission of a dark coloured urine, kidney necrosis being secondary to the haemoglobinuria.

TABLE 3

EVOLUTION OF THE CLINICAL SYMPTOMS OF COPPER INTOXICATION

	Blood	Liver	Kidney	Other organs
Accumulation phase	Transaminases↑ Lactic dehydrogenase↑	Copper↑		
Haemolytic crisis	Copper ↑ Methaemoglobin ↑ Heinz bodies in erythrocytes CPK↑ Hemoglobin↓ Glutathion↓	Liver necrosis	Haemoglobinuric nephrosis (swollen, enlarged and dark)	Brown to dark icteric (yellow) colour of organs and fat

Diagnosis

Haemolytic icterus is due mainly to a copper toxicity in sheep. Copper toxicity should be confirmed by the analysis of copper in the diet and more specifically by the determination of liver copper concentrations. Toxic levels of copper in liver is characterised by copper concentrations of over 1 000 mg/kg DM (Table 4). During the haemolytic crisis plasma copper concentrations attain values above 130 μg/100 ml. High plasma copper concentrations should however be distinguished from any infectious diseases or inflammatory processes, since MacCosker (1968) has shown that any inflammation, turpentine injection for example, would double plasma copper concentration.

If the level of molybdenum in the diet is low (< 0.2 ppm), then a dietary level of 15 mg Cu/kg DM should be considered as an upper limit beyond which a potential copper toxicity would be imminent. If, however, the dietary level of molybdenum is higher than 0.2 ppm, then 20 mg Cu/kg DM should be adopted as the upper limit for sheep.

TABLE 4

DIAGNOSIS OF COPPER TOXICOSIS IN SHEEP

Sample	Parameter	Limit	Comments
Diet	Copper	> 15 mg/kg DM	
	Molybdenum	< 0.2 mg/kg DM	
Blood	Haematocrit	< 30 %	
	Coloured plasma	–	Brown
	Plasma copper	130 µg/100 ml	To be distinguished from infectious or inflammation disease
Urine	Colour	–	Brown to dark
Liver	Copper	1 000 mg/kg DM	

Treatment

Before the haemolytic crisis is manifested, it is possible
to reduce substantially the copper intake of the affected sheep
by grazing copper deficient pastures or by feeding forage,
deficient in copper. Liver copper stores may be depleted by
the supplementation of the diet with molybdenum and sulphur.
The daily addition of 13 mg of molybdenum, as ammonium
molybdate, and 3 g of sulphur per kg of DM to the diet for a
period of 12 weeks may be administered without any detrimental
effect (Lamand et al.,1980).

According to Bremner et al. (1976) zinc supplementation
(diet containing up to 420 mg Zn/kg DM) may provide sufficient
protection against copper toxicosis.

Any specific treatment of copper toxicity in sheep should
be completed with treatments of secondary insufficiency of
liver and kidney function.

REFERENCES

Bremner, I., Young, B.W. and Mills, C.F., 1976. Protective effect of zinc
   supplementation against copper toxicosis in sheep. Brit. J. Nutr.
   36, 551-561.

Clarke, N.J. and Laurie, S.H., 1980. The copper-molybdenum antagonism in
   ruminants. I. Formation of thiomolybdates in animal rumen. J. Inorg.
   Biochem. 12, 37-43.

Coppenet, M. 1980. Personnal communication.

MacCosker, P.J., 1968. Observation on blood in the sheep. I. Normal copper
   status and variations induced by different conditions. Res. Vet.
   Sci. 9, 91-101.

Lamand, M., Lab, C., Tressol, J.C. and Mason, J.M., 1980. Biochemical
   parameters useful for the diagnosis of mild molybdenosis in sheep.
   Ann. Rech. Vet. 11, 141-145.

Mason, J.M., Lamand, M. and Kelleher, C.A., 1980 (in press). Duodenal
   infusion of ^{99}Mo labelled sodium di- and trithiomolybdates in sheep;
   the effects on plasma copper and on the diamine oxidase activity of
   caeruloplasmin (Ec. 1.16.3.1.) Brit. J. Nutr.

Soli, N.I., 1980. Chronic copper poisoning in sheep, a review of the
   literature. Nord. Vet. Med. 32, 75-89.

Suttle, N.F. and Price, J., 1976. The potential toxicity of copper rich
   animal excreta to sheep. Anim. Prod. 23, 233-241.

DISCUSSION

J.B. Ludvigsen (Denmark)

It is very interesting to see that your creatine phosphokinase is increasing too. That would seem to indicate that
something is happening in the skeletal musculature. Also the
myoglobin may well be heavily affected. Can you elaborate a
little on this?

M. Lamand (France)

I have never seen anything in the literature on these
points. However, creatine phosphokinase is supposed to rise
because of membrane rupture of cells, probably muscle cells.
I would say copper invasion increases strongly the oxidation
state and the oxidation of membranes. This oxidation state
has been studied in erythrocytes.

J.B. Ludvigsen

Yes, this is true. The molecular weight of myoglobin is
one quarter that of haemoglobin; they have, so to speak, the
same configuration. So the myoglobin could also be heavily
affected, that being the primary cause of the degradation of
the muscular cells and the cell membranes; because of a lack
of oxygen metabolism in the musculature they are in an
anaerobic state.

M. Lamand

But remember that creatine phosphokinase is liberated
out of the cell earlier than myoglobin because of the difference in molecular weight.

K.L. Robinson (UK)

I would like to raise the question of molybdenum. I am
surprised to hear that a low level of molybdenum is an overall
problem in France for one has not heard of this as a problem
in, say, adjacent parts of England. I wonder if any of the UK

soil scientists could comment on this point.

M. Lamand

We were surprised too. We expected higher molybdenum
levels in some places but we did not find them.

R.J. Unwin   (UK)

Grass and hay is usually only analysed for molybdenum by
ADAS in England and Wales when there is reason to suspect that
high molybdenum may be inducing copper deficiency. Of 3 000
samples analysed between 1968 and 1979, 86% were found to
contain less than 2.0 mg/kg molybdenum in the dry matter.
Because of the very biased nature of the samples tested, the
proportion of the total grassland area which has less than this
molybdenum content will, in fact, be much higher.

G.A. Fleming   (Ireland)

I think it is generally accepted that high molybdenum
regions are more common in Ireland and in some parts of the UK
than they are in France. There are sound geochemical reasons
for this. However, I would like to comment on Dr. Lamand's
reference to levels of molybdenum in hay. We may be on
dangerous ground here because, as far as I can recall, in the
classical work of Ferguson, Lewis and Watson in the early
1940s, whilst the molybdenum content of hay, or mature grass,
was lower than that in fresh grass, the water-soluble molybdenum was higher. So, a low figure in hay does not necessarily
mean a low availability to the animal.

M. Lamand

Generally young grass is richer in trace elements and
also in molybdenum, but we used hay from natural pasture, which
is a very common feed in France, as a reference. However,
molybdenum levels in younger grass are not very high, except
perhaps in very young grass. We also noticed that nitrogen
fertilisation can modify molybdenum content of grass to a
considerable extent; the more nitrogen fertiliser applied, the

more the decrease in molybdenum level. So the situation can
be completely modified by economic factors.

G.A. Fleming

I think it is quite difficult to find high molybdenum
herbage on a freely drained soil. Certainly in high pH soils
molybdenum tends to become more available, but where soils are
poorly drained it has been well documented that molybdenum
markedly accumulates in the plant.

M. Lamand

This is well known but we do not have many such soils in
France.

M.D. Webber (Canada)

What is considered to be a sufficient level of molybdenum in forage?

M. Lamand

I would say that about 0.5 ppm would be much better than
0.1 ppm. I feel that an increased molybdenum level would
extend the limit for copper toxicity and would thus facilitate
using copper for sheep. But molybdenum itself is rather
dangerous, and it is difficult to use one element as an antagonist against another element because this introduces a
complicated system which can get out of hand eventually.

D.B.R. Poole (Ireland)

At the last EAAP meeting in Munich, Dr. Enka from Jena
produced experimental results suggesting a molybdenum requirement by the ruminant (I think it was the goat he was referring
to) of something like 0.2 ppm.

J.B. Ludvigsen

This is just a theoretical question. When you state
that molybdenum in hay is low, what are you comparing it to?

You may say that it is low compared to the requirements of animals. But there are so many other factors involved, fertiliser practice, dietary levels of copper and so on. From times past, when no fertilisation was used, is there a balance between molybdenum and copper? It may be the case for grass of various sorts in various countries that there is a concentration of molybdenum which is characteristic but which is low compared to what the animals require.

M. Lamand

I think it is truly the balance between molybdenum and copper. I have observed molybdenum toxicity in the vicinity of a steel works; also, I have observed situations varying from 1 ppm to 20 ppm molybdenum where animals can be managed only if dietary copper levels are raised; you have something like two parallel curves but it is difficult to quantify exactly.

J.B. Ludvigsen

But the concentration of molybdenum will be generally characteristic for the species of grass or hay. When you term it low this is in relation to what the animals need. It is not low from a plant species point of view.

M. Lamand

Yes, I agree.

IMPLICATIONS OF APPLYING COPPER RICH PIG SLURRY TO GRASSLAND - EFFECTS ON THE HEALTH OF GRAZING SHEEP

D.B.R. Poole

The Agricultural Institute, Dunsinea, Castleknock,
Co. Dublin, Ireland.

ABSTRACT

A series of experiments involving sheep has been undertaken to assess the degree of risk to sheep from the use of copper rich pig slurry. Copper from slurry was shown to be absorbed by sheep, although at a lower availability than from copper sulphate. Availability was further reduced by the presence of soil and other minerals in the diet. While a potential hazard to sheep was demonstrated if slurry contaminated the pasture, no hazard was identified on pastures which had previously received heavy slurry applications.

The overall use of copper in pig feeding in Ireland and the problems and benefits of proper disposal of slurry have been discussed. The acceptable disposal of copper rich slurry on grassland should be possible within 0.6 km radius of each pig fattening unit.

INTRODUCTION

An interdisciplinary study of the possible effects of the use of copper rich pig slurry on grassland has been undertaken. Preliminary results were given in an earlier paper (Poole and McGrath, 1978) and more complete results later (McGrath et al., 1980). The effects on plants and soils have already been described in this meeting (McGrath). In this meeting, I wish (a) to review our previous findings, (b) to report some further work, and then to attempt a realistic assessment of the benefits and risks arising from copper rich pig slurry in Ireland.

MATERIALS AND METHODS

The soils and experimental procedures have been previously described (McGrath et al., 1980). On one loam soil (Soil No. 3 - Hoarstone) following slurry applications at 0, 23, 46 and 115 m^3/ha three times annually for three years, in the fourth year five sheep were rotationally grazed on each treatment. No slurry was applied. Nitrogen was applied at 170 kg/ha. Liver samples were obtained before the commencement, by biopsy, and at the termination following slaughter. Herbage samples were collected prior to grazing each plot. Samples of fresh faeces were collected from the ground following each grazing. Four grazing cycles were completed within a 15 week period between June and October.

RESULTS

(a) Review of earlier results

In our earlier report (McGrath et al., 1980), we described a grazing experiment in which the paddocks received applications of copper rich pig slurry at 20 m^3/ha three times during each of two grazing seasons. This resulted in a marked increase in faecal copper; liver copper levels were signif- icantly increased (by ca 300 - 150 mg/kg dry matter respectively for loamy sand and loam soils). Although the levels of liver copper did not reach those associated with chronic copper poisoning, this was a clear indication of a potentially dangerous trend.

In the absence of applied slurry, on the loam soil, sheep at a low stocking rate accumulated more copper than those on the high stocking rate. This increase was of similar degree to that resulting from slurry application at the high stocking rate. It was not associated with increased herbage copper as indicated by faecal copper assay. This effect of stocking rate was not seen on the sandy loam soil (Poole and McGrath - unpublished).

In a series of feeding experiments involving housed sheep, we showed that the copper in dried copper rich slurry (1 080 Cu mg/kg dry matter) was absorbed by sheep and stored in the liver. The availability was much less than that of copper as copper sulphate at similar levels, but greater than that of copper sulphide. The availability of copper, supplied as slurry copper, was reduced by the addition of soil (2½% and 10% additions).

We also reported on the inclusion in the feed of copper (as copper sulphate) in combination with the other mineral components of slurry. While this reduced the availability of copper, it did not reduce it to that of slurry copper. We therefore concluded that the relatively poor absorption by sheep of copper from slurry is due in part to its chemical form, as well as to the adverse interaction with other slurry components.

(b)   New results

In our recent work our intention has been to examine the effects of previous slurry applications, where direct pasture contamination can be excluded.

McGrath has already, in this meeting, given the results of copper in soil and mat following slurry applications. Soil copper increased by 30 mg/kg in the top 5 cm and by 5 mg/kg in the 5 - 10 cm layer. The results of copper and zinc analyses (Table 1) show only a trend towards higher copper levels in herbage associated with slurry use. Increases in herbage zinc also occurred. These effects are also reflected in the faecal analysis results. The acid insoluble ash values indicated mean soil intakes below 5% of total dry matter intake. Comparisons between animals on rotational and set stocked grazing systems are of uncertain value, but it appears that the soil intakes in this case are similar to those previously recorded at low stocking rate.

TABLE 1

SEASONAL MEAN VALUES FOR COPPER AND ZINC IN HERBAGE AND COPPER, ZINC AND ACID INSOLUBLE ASH IN FAECES

| | Herbage | | Faeces | | |
	Copper*	Zinc*	Copper*	Zinc*	Ash**
Control	10.3	42.3	20.8	126.4	73
Low slurry	11.7	49.6	23.3	138.0	81
Medium slurry	12.7	48.2	21.0	137.6	74
High slurry	12.1	60.1	25.4	151.8	84

* mg/kg   ** g/kg

The average increases in copper, zinc and iron in liver are given in Table 2. These values show no statistically significant effect resulting from slurry treatment. The trend is for accumulation of copper to be reduced by slurry application. It is reasonable to suggest that this could be associated with the increased values for zinc and possibly iron in herbage.

TABLE 2

INCREASES IN LIVER COPPER, ZINC AND IRON LEVEL AFTER 15 WEEKS (mg/kg)

Treatment	Copper	Zinc	Iron
Control	224	-47	251
Low slurry	166	2	109
Medium slurry	151	-20	101
High slurry	176	3	160

DISCUSSION

The results of the programme of research which have been
summarised in this paper indicate that copper in copper rich
pig slurry represents a hazard to sheep, and probably to other
susceptible animals, if it directly contaminates the food, i.e.
the pasture. This is most likely to occur if the slurry is
spread on the pasture before or during grazing, but it can also
be incorporated during harvesting (Dalgarno and Mills, 1975).
This finding is in broad agreement with the results of a number
of other studies (Batey et al., 1972; Dalgarno and Mills, 1975;
Price and Suttle, 1975, etc.).

Apart from this direct effect, copper could reach the
animal by plant uptake or by the ingestion of soil by the
grazing animal following long-term slurry applications. The
results presented at this meeting by McGrath suggest that
copper accumulation by pasture species should not present a
hazard to grazing animals. The feeding experiments confirmed
the adverse effect on the absorption of copper by soil itself
and by the other mineral components of slurry, leading us to
the conclusion that copper toxicity due to soil intake is
unlikely to occur. This conclusion is supported by the results
presented in this paper.

Chronic copper poisoning in sheep has recently been
comprehensively reviewed (Søli, 1980). The risk of toxicity

is governed by a number of factors, i.e. the susceptibility of the animal (species, breed and age), the chemical form and level of copper in the feed, its composition and the period of intake. The availability of copper was shown in this study and by Suttle et al., 1975, to be reduced by the inclusion of soil in the diet. We have shown soil intakes comparable with that reported by other workers (Field and Purves, 1964; Healy, 1967; Nolan and Black, 1970). The inclusion of certain other mineral elements in the diet also reduces copper absorption by sheep and cattle, particularly calcium, cadmium, zinc, iron and molybdenum (Underwood, 1977). While molybdenum and cadmium should not occur to any extent in pig slurry, values for calcium, zinc and iron can be high. We have shown that their inclusion in an artificial situation reduced copper absorption considerably. Nevertheless, the copper in dried pig slurry was more poorly absorbed by sheep than was a soluble salt of copper.

In Ireland, copper deficiency affecting grazing animals is widespread. Soils derived from black shales of the Namurian period are often rich in molybdenum and form a significant proportion of our grassland area. Induced copper deficiency appears to be one of the most frequent and important nutritional problems affecting young beef cattle; it also affects many of our dairy herds. It can cause general ill-health, poor growth rate, reduced milk yield and impaired breeding performance (Poole, 1973). In sheep, the incidence of swayback varies from year to year, but can be significant and widespread, although it would not be classed as a serious national problem.

We have undertaken a number of surveys of bovine copper status, based on blood analysis. In one such survey of herds in Co. Clare, in which about 600 herds were involved, 70% of the herds were of a low copper status. Similar results have been obtained for other smaller scale surveys in widely scattered parts of Ireland.

While much of this hypocupraemia is the result of anomalous molybdenum levels, it is also associated with pasture copper levels which are frequently below 6 mg/kg. It is therefore reasonable to consider the beneficial effects of copper rich pig slurry as well as its disadvantage. In this connection Todd 1978 has, for Northern Ireland, related the reduction of copper usage in agriculture with the increased incidence of bovine copper deficiency. He estimated that from 1900 to 1950, 3 000 tonnes of copper sulphate would have been used annually in Northern Ireland, mainly as a potato spray, whereas in recent years only 300 tonnes of copper sulphate were used, mainly in pig feeds.

The value of supplementary copper in pig feeding has already been outlined in this meeting. In Ireland, it is estimated that 300 000 tonnes of pig fattening feed is produced annually and that approximately 80% of this is supplemented with copper to about 200 mg/kg (Hanrahan, T. personal communication). This therefore involves the annual disposal of 180 tonnes of copper sulphate. Various levels of application of copper have been suggested for grassland - if the figure of 2.5 kg/ha is accepted, the Irish pig industry would require 18 000 ha for slurry spreading.

In Ireland, 93% of our fattening pigs are reared in herds of 100 pig places and above (Anon, 1980) of which there are 400. Allowing for 2.5 kg copper/ha, the average land area required per farm for slurry disposal would be 42 ha. When one looks at the large fattening units (> 1 000 pig places), of which we have 100 accounting for 63% of our fattening pigs, the land area required for each such farm for copper disposal would be 114 ha. If this were available in a complete circle about the farm, the greatest distance for slurry disposal would be 0.6 km. I therefore cannot accept that we face a national problem regarding slurry copper disposal.

On the other hand, if you look at the situation regarding the phosphorus content of this slurry, by spreading at 2.5 kg

copper/ha 83 kg phosphorus/ha would be applied. This appears
to me to be a much more serious situation.

In closing therefore, I would like to refer to the guidelines previously published (Anon, 1978) and in particular to
No. 3. This states that utilisation of slurry should be
related to plant nutrient requirements. If this is conformed
to in relation to phosphorus, the annual application of copper
should not exceed 1 kg/ha.

REFERENCES

Anon, 1978. In: W.R. Kelly (Editor) Animal and Human Health Hazards
    Associated with the Utilisation of Animal Effluents. Commission
    of the European Communities, Luxembourg p 302-304.

Anon, 1980. Estimated numbers and percentage distributions of
    (a) holdings with cattle and pigs and (b) cattle and pigs returned,
    classified by the number in the holding, Dec., 1979. Central
    Statistics Office, Dublin.

Batey, T., Berryman, C. and Line, C., 1972. The disposal of copper-
    enriched pig manure slurry on grassland. J. Br. Grassld. Soc.
    27, 139-143.

Dalgarno, A.C. and Mills, C.F., 1975. Retention by sheep of copper from
    aerobic digests of pig faecal slurry. J. agric. Sci., Camb. 85,
    11-18.

Field, A.C. and Purves, D., 1964. The intake of soil by grazing sheep.
    Proc. Nutr. Soc. 23, XXIV-V.

Healy, W.B., 1967. Ingestion of soil by sheep. Proc. New Zealand Soc.
    Anim. Prod. 27, 109-120.

McGrath, D., Poole, D.B.R. and Fleming, G.A., 1980. Hazards arising
    from the application to grassland of copper rich pig faecal slurry.
    In: J.K.R. Gasser (Editor) Effluents from Livestock. Applied
    Science Publishers Ltd., London p 420-431.

Nolan, T. and Black, W.J.M., 1970. Effect of stocking rate on tooth wear
    in ewes. Ir. J. agric. Res. 9, 187-196.

Poole, D.B.R., 1973. Thesis: Studies on induced copper deficiency in
    cattle. Dublin University.

Poole, D.B.R. and McGrath, D., 1978. Animal effluent management and the
    survival and persistence of priority contaminants - Toxicological
    agents. In: W.R. Kelly (Editor) Animal and Human Health Hazards
    associated with the Utilisation of Animal Effluents. Commission of
    the European Communities, Luxembourg p 302-304.

Price, J. and Suttle, N.F., 1975. The availability to sheep of copper in
    pig-slurry and slurry-dressed herbage. Proc. Nutr. Soc. 34, 9A-11A.

Søli, N.E., 1980. Chronic copper poisoning in sheep. Nord. Vet-Med.,
    32, 75-89.

Suttle, N.F., Alloway, B.J. and Thornton, I., 1975. An effect of soil
    ingestion on the utilisation of dietary copper by sheep. J. agric.
    Sci., Camb. 84, 249-254.

Todd, J.R., 1978. The copper status of ruminant animals in Northern
    Ireland in relation to the usage of copper compounds in agriculture.
    In: M. Kirchgessner (Editor) Proc. 3rd. Int. Symp. Trace Element
    Metabolism in Man and Animals, Freising-Weihenstephan, p.486-489.

Underwood, E.J., 1977. Trace elements in human and animal nutrition.
    Academic Press, London, 71-72.

DISCUSSION

P.H.T. Beckett *(UK)*

    Do you anticipate any problems from the phosphorus?
Usually we are worried about phosphorus fixation. Are the
surpluses you describe likely to lead to any difficulties?

D.B.R. Poole *(Ireland)*

    There are two aspects here that could be commented on.
The one on which I do not feel qualified to comment is where
a man is putting out a given amount of phosphorus, which is
contained in a given amount of slurry with all its other
components. I think this will have an effect on soil structure
and a number of other things but I do not intend to go into that
area at all. Being more specific in relation to phosphorus,
again it is very hard to separate this because when one has seen
cases of farms where phosphorus levels are high, obviously
other things have been high too. However, I do associate general,
non-specific problems of production, particularly in the sheep,
with soil levels of phosphorus in excess of 20 ppm. I have no
evidence for this other than to say that the coincidence of
problems and high levels of phosphorus in cases I have dealt
with appears to be too high to be just coincidental.

R. Braude *(UK)*

    I am puzzled by your figures on the two stocking rates at
these very high levels of copper. Have you any explanation for
that?

D.B.R. Poole

    First of all, the two stocking rates were both without
additional copper. We have evidence to suggest that the copper
content of the feed the sheep were on was similar: the faecal
copper levels of the sheep at the two stocking rates were similar,
so it was not a case of different copper levels in the feed.
There are two possible reasons why it may have occurred. One is
that the animals may have been eating more grass at the low

stocking rate, and therefore, their copper intake would have
been increased. I tend to discount that theory because it only
happened on the loam soil and not on the sandy soil. The other
possibility which I tend to favour is that the soil intake, by
the grazing animal, is related to stocking rate. We have shown
this clearly and others have shown it also. Therefore, the soil
intake at the high stocking rate would have been much higher
and would have interfered more severely with copper absorption
by the animal.

A. Dam Kofoed *(Denmark)*

   I think it is a very fair statement you have made,
Dr. Poole. If we utilise pig slurry according to the phosphorus
content there will be no problems with copper.

R.J. Unwin *(UK)*

   Obviously there are others in the room apart from me who
took part in the workshop at Haren earlier in the year when the
accumulation of phosphorus from slurry was the topic of the
meeting. There may be people here who would be interested to
see the Proceedings of that meeting when it is published. I
would just make the point that I would accept that 83 kg of P
is towards the maximum level but if we are talking about grass
which is going to be cut for forage production then, in an organic
form, I do not think that figure of 83 would be excessive if
you were cutting several times during the season. Neither, of
course, would it be excessive for certain arable crops if one
was considering something like potatoes.

A. Madsen *(Denmark)*

   Dr. Poole, in your slides you showed that if you have
more than 1 000 pigs to a farm you need only 140 ha. This seems
to be a fairly big farm, at least compared with Danish conditions.
You also gave a figure that you need only to take a maximum
journey of 0.6 km. But if you don't have 140 ha on that swine
production unit it means that you have to hope that your neighbour will buy all the slurry. If this is not the case then

you can easily need to travel 6 or even 60 km. What do you
do then?

D.B.R. Poole

I think to jump from half a kilometre to 60 kilometres
is exaggerating the situation. I take your point entirely
that I am putting up a false situation; I am doing it merely
to illustrate a point. If you take the farmer as being the centre
of a circle and take a radius from that point of 0.6 km, this
gives you the 140 ha. If you assume that half the farmers in
that area refuse to co-operate and you need to double the area,
you are only coming out to a radius of something like 0.7 km.
This is the sort of situation I am talking about.

The other point which is more relevant than the cooperation of the farmers is the coincidence of other pig units.
If you have two pig units close to each other then you are
affecting the whole argument. However, I put this up merely
to try to put the thing into perspective because there is a
tendency to argue that a large pig unit has a massive haulage
problem with slurry. I would tend to counter that argument by
saying that the farmer who is getting the benefit of copper-
containing feed for his pigs does not have that benefit offset
by transport costs.

A. Dam Kofoed

Do you have any difficulty in getting the neighbouring
farmers to take the slurry? Do they pay for it?

D.B.R. Poole

It depends on the neighbour, I think. In Ireland we have
a relatively spread-out pig industry. We have only one small
county in Ireland where the number of pig units would probably
make this argument quite difficult to maintain. In County
Cavan there are quite a lot of small sized (100 - 200) pig units
and I think there would be trouble there in disposing of the
slurry to neighbours because they would probably have pigs of

their own. Apart from that, I would not expect much trouble. In some areas in Ireland it is actually paid for, in other areas it is taken for the haulage cost. In no case that I know of does the pig producer have to pay for the slurry to be disposed of. He is either giving it away free or he is getting some fairly nominal return for it.

J.H. Voorburg (Netherlands)

Just a comment: you accept 2.5 kg/ha/year of copper; this means that you accept a layer of 7 cm on grassland, and increase of 3 ppm per year. So after 100 years you will have quite a high level of copper.

D.B.R. Poole

I would point out first that I am putting up 2.5 kg as a position of argument; I am not proposing it as a safe level. However, I would ask you, do you expect that in the year 2080 pig farmers will be feeding pigs in exactly the same way as they are doing it today - because I don't.

D. McGrath (Ireland)

Dr. Voorburg's comment is presuming that all the pig slurry is going on the same land. It so happens in Ireland that there is a lot of land to go around. If you do David Poole's calculations in another way and spread all of the slurry over the total surface of the land that is in agricultural usage, you would be adding 10 g of copper per hectare per year. That is less than is actually lost by leaching.

EFFECT OF COPPER SUPPLIED IN THE FORM OF DIFFERENT Cu-SATURATED SLUDGE SAMPLES AND COPPER SALTS ON THE Cu-CONCENTRATION AND DRY MATTER YIELD OF CORN GROWN IN SAND

S. Gupta and H. Haeni

Swiss Federal Research Station for Agricultural Chemistry
and Hygiene of Environment,
CH-3097 Liebefeld-Bern, Switzerland.

ABSTRACT

A growth chamber experiment (sand culture) was conducted in order to investigate the effect of degree of copper saturation of a sludge sample on the dry matter yield of corn and on copper concentration. Various application levels of copper in the range of 2.5 to 40.8 µg Cu/g sand were also tested.

The results have clearly shown that the increasing degree of copper saturation of the sludge sample, from 9.8 to 98.3%, caused an increasing toxic effect on the corn growth and on the enhancement of copper concentration. This observation is true for all the application levels, however, the effects are more pronounced at higher levels, i.e. 20.4 and 40.8 µg Cu/g sand. Sludge which was 3.4% copper saturated (containing 2 077 µg Cu/g dry sludge) did not affect, at all the application levels, either growth or copper concentration. The limiting value of copper in sludge, in Switzerland and many other countries, is 1 000 µg Cu/g dry sludge.

INTRODUCTION

Sewage sludges not only supply plant nutrients (like nitrogen, phosphorus, calcium, magnesium etc.) but also a range of growth-essential and non-essential metals to soils. The metals, if applied to the soil at excessive rates can produce injurious or toxic effects on either crops or on consumers. The elements like Cu, Zn, Cd, Ni, Cr, Pb, Mo, As and B are present in sludges at much higher levels than are found in soils, and the concentrations present are variable. The problem of metal content in sludge produced in rural areas, where there is little or no industry within the catchment areas of waste water treatment plants is significantly less than in industrial areas. It is clear that the application of sewage sludges specially loaded with metals to cultivated land will lead to substantial metal accumulation in soils, which in long term could adversely affect the soil fertility. In order to avoid serious pollution of soils, which could result from careless utilisation of metal loaded sludges, rational standards defining maximum tolerable levels of potentially toxic heavy metals in sludge need to be defined. (Furrer, 1977; Furrer et al., 1980).

In Switzerland, standards defining the maximum limit of potentially toxic heavy metals in sludge were established in 1975 (Keller, 1976; Keller, 1977a, b) and have been used continuously in sludge control and recommendations since 1977 (Furrer et al., 1980). These values are quite similar to values proposed in the Federal Republic of Germany (Kloke, 1980). In working out these values for sludges, the principle used was that the continuous use (at 5 tonnes of dry sludge/ha/year) of sludge for a period of at least 50 years should not increase the total metal concentration of the upper 20 cm layer/ha of a normal soil (ca. 2 500 tonnes of dry soil) to a level higher than the limiting values of soil. However, as far as can be ascertained, the validity of these values have not so far been tested and verified through growth experiments.

An important implication of our basic studies (Marinsky et al., 1980) on Cu binding by organic matter is that as the proportion of its maximum binding capacity filled by a metal increases, the relative strength of binding decreases and, hence, the relative plant availability of the Cu should increase. In simple words, sludge organic matter which is saturated to a lower percentage with Cu (as a percentage of its total exchangeable acidity) would have less easily exchangeable copper compared to the same sludge saturated to a higher percentage by Cu. This points out the danger that application of sludges which have a high percentage of metal saturation would not only add to a soil a greater quantity of metals in a short time but, also a large proportion of lightly bound metal, and would result in early exhaustion of metal binding capacities of soils.

One growth experiment which has a practical application, was conducted to test the hypotheses presented earlier; it was based on the results of model adsorption experiment (Marinsky et al., 1980). The results of such an experiment would give us the necessary basic information needed for the existing tolerance limits and may help in improving them as well.

In this paper results of the experiment are presented and discussed with the following precise objectives in mind:

1. To find the relationship between the different percentages of Cu saturation of the same sludge sample, and the dry matter yield of corn.

2. To compare the effects of copper applied in forms of copper salts and copper-sludge samples (Cu-SS-) saturated through Cu to different percentages of total acidity of the original sample, on the dry matter yield of corn and copper concentration.

MATERIALS AND METHODS

A sludge sample was collected from a waste water treatment plant having a facility of dewatering through dry beds. A bulk sample of dewatered sludge was dried in the oven at $105^{\circ}C$. It was analysed for nutrient elements i.e. N, P, K, Ca and Mg and heavy metals. These results are presented in Table 1.

TABLE 1

CHEMICAL COMPOSITION OF SEWAGE SLUDGE (SS) USED IN THE SAND CULTURE GROWTH EXPERIMENT

Factor		Content	Unit
Dry matter		98.1	%
Organic matter		29.1	%
Ash		69.0	%
Nitrogen	(N)	1.6	%
	NH_4-N	0.11	%
Phosphate	(P)	0.64	%
Potassium	(K)	0.30	%
Calcium	(Ca)	6.9	%
Magnesium	(Mg)	0.7	%
Cadmium	(Cd)	5	µg/g
Nickel	(Ni)	31	µg/g
Chrome	(Cr)	58	µg/g
Copper	(Cu)	161	µg/g
Zinc	(Zn)	1 225	µg/g
Lead	(Pb)	256	µg/g
pH		6.9	
Total exchangeable acidity		1.8	meq/ DM

The data in Table 1 indicate that the present sample is low in heavy metals, especially copper content, which is 161 µg/g sludge. The total exchangeable acidity which includes most of COOH and OH groups, is measured by the method suggested by Schnitzer and Gupta (1972). In this procedure, the barium is used to replace the hydrogen ions and the replaced H ions are potentiometric titrated to pH 8.4 with the help of 0.5 N HCl. The total exchangeable acidity of the sludge sample is 1.8 meq/g dry sludge. This value is taken as 100% saturation, because the maximum amount of any metal which could be absorbed on a sludge sample would be equal to 1.8 meq/g sludge. In order to prepare four copper containing samples saturated to different degrees, four original dried sludge samples, re-dissolved in water, are mixed with the desired amounts of copper (meq) in the form of copper nitrate. The mixtures are shaken separately for a period of two days. The mixed samples are dried in an oven at 105°C. The degrees of saturation obtained in this way are 3.6, 9.8, 34.6 and 98.3% which are abbreviated as Cu-SS-3.4%, Cu-SS-9.8%, Cu-SS-34.6% and Cu-SS-98.3% respectively. The remaining experimental conditions are summarised in Table 2.

TABLE 2

EXPERIMENTAL CONDITIONS OF SAND CULTURE GROWTH EXPERIMENT

Copper sources	% Cu-saturation based on total-acidity	Abbreviation	Actual Cu content mg/g sludge
a. $Cu(NO_3)_2$	–	Cu-S	–
b. Sludge	3.6	Cu-SS-3.6	2.07
c. Sludge	9.8	Cu-SS-9.8	5.6
d. Sludge	34.6	Cu-SS-34.6	19.8
e. Sludge	98.3	Cu-SS-98.3	56.2

Substrate	:	Quartz sand (10 kg/pot)
Plant	:	Corn (silage) LG 11
Cu-Doses	:	2.5, 5.1, 10.2, 20.4, 40.8 µg Cu/g sand
Fertilisers	:	N, P, K, Mg, S as salts and micronutrients as solution
Replication	:	3

RESULTS AND DISCUSSION

In the sand culture growth experiment, four copper sewage sludge complexes (Cu-SS-3.6%, Cu-SS-9.8%, Cu-SS-34.6% and Cu-SS-98.3%) representing four degrees of copper saturation with respect to total exchangeable acidity of the original sludge sample, and copper nitrate, were used as sources of copper.

Effect of degree of Cu saturation of sludge on the dry matter yield of corn

The dry matter yields of straw, cob, grain and root are presented in Table 3. The total dry matter yield (i.e. of straw + cob + grain + root) is graphically presented in Figures 1 and 2. A close study of Table 3 and Figures 1 and 2 suggests the following:

- The dry matter yield of straw, cob, grain and root and the total dry matter yield decreased significantly as the dose of copper applied in the form of copper nitrate increased.

- At all levels of copper application, the copper sludge complexes, i.e. Cu-SS-9.8%, Cu-SS-34.6% and Cu-SS-98.3%, affected the total dry matter yield of corn. However, toxic effects on corn growth were more pronounced at copper applications of 20.4 and 40.8 µg Cu/g sand and the results were statistically significant.

- A significant decrease in dry matter yield was recorded as the degree of copper saturation of sludge increased and this tendency was true for all levels of application.

- In contrast, the very weakly saturated copper sludge complex (Cu-SS-3.4%) did not affect the total dry matter yield at lower applications, but at higher applications a slight increase in the yield was recorded.

It seems that the dry matter yield of corn is affected by two factors: degree of copper saturation of the sludge

TABLE 3

EFFECT OF DEGREE OF COPPER SATURATION OF SLUDGE (%) AND COPPER DOSES (µg/g SAND) ON THE DRY MATTER YIELD AND COPPER CONCENTRATION OF CORN (STRAW, COB, GRAIN AND ROOT) IN GROWTH CHAMBER EXPERIMENT (SAND CULTURE). (RESULTS ARE AVERAGE OF THREE REPLICATIONS)

Copper doses µg/g sand	Copper salt Cu(NO$_2$)$_3$		Copper-sludge complexes (Cu-SS), saturated to following %							
			Cu-SS-3.6%		Cu-SS-9.8%		Cu-SS-34.6%		Cu-SS-98.3%	
	DM	Conc.	DM	Conc.	DM	Conc.	DM	Conc.	DM	Conc.
STRAW										
2.5	203	14	184	11	172	13	180	14	180	12
5.1	166	18	160	13	174	15	171	17	168	21
10.2	155	24	164	12	175	14	169	18	155	21
20.4	90	26	173	11	164	17	105	26	96	34
40.8	31	32	203	11	116	25	24	37	8	34
COB										
2.5	33	6	31	5	36	6	30	7	31	7
5.1	29	8	25	6	34	7	35	7	38	7
10.2	33	8	29	6	30	8	33	8	30	8
20.4	17	10	27	6	24	9	27	10	25	9
40.8	0	0	36	6	23	10	0	0	0	0
GRAIN										
2.5	156	4	160	4	148	4	145	3	133	4
5.1	141	4	115	3	154	4	144	4	158	4
10.2	159	5	146	3	143	4	130	4	139	5
20.4	26	7	146	5	114	5	89	5	30	7
40.8	0	0	156	4	113	5	0	0	0	0

TABLE 3 (CONTINUED)

Copper doses μg/g sand	Copper salt Cu(NO₂)₃		Copper-sludge complexes (Cu-SS), saturated to following %							
			Cu-SS-3.6%		Cu-SS-9.8%		Cu-SS-34.6%		Cu-SS-98.3%	
	DM	Conc.	DM	Conc.	DM	Conc.	DM	Conc.	DM	Conc.
ROOT										
2.5	52	90	40	37	27	60	35	62	38	72
5.1	23	138	24	58	33	114	32	157	32	159
10.2	31	282	34	110	28	171	35	356	25	303
20.4	18	529	32	134	26	258	17	489	20	473
40.8	8	724	44	148	13	492	3	0	2	957
STRAW + COB + GRAIN										
2.5	392	8	374	7	346	8	355	8	345	8
5.1	336	10	301	7	362	8	350	9	360	11
10.2	347	12	339	7	348	9	331	10	323	11
20.4	134	14	346	7	301	10	221	13	150	17
40.8	31	32	396	7	252	13	24	37	8	34
STRAW + COB + GRAIN + ROOT										
2.5	443	29	414	14	373	21	390	21	382	24
5.1	359	42	324	20	395	31	382	46	396	48
10.2	378	80	373	33	376	49	366	97	348	84
20.4	152	143	378	39	327	72	237	132	170	131
40.8	39	189	440	42	265	133	27	-	10	496

COPPER SOURCES

DM : Dry matter yield (g/pot)
Conc : Copper concentration (μg/g)

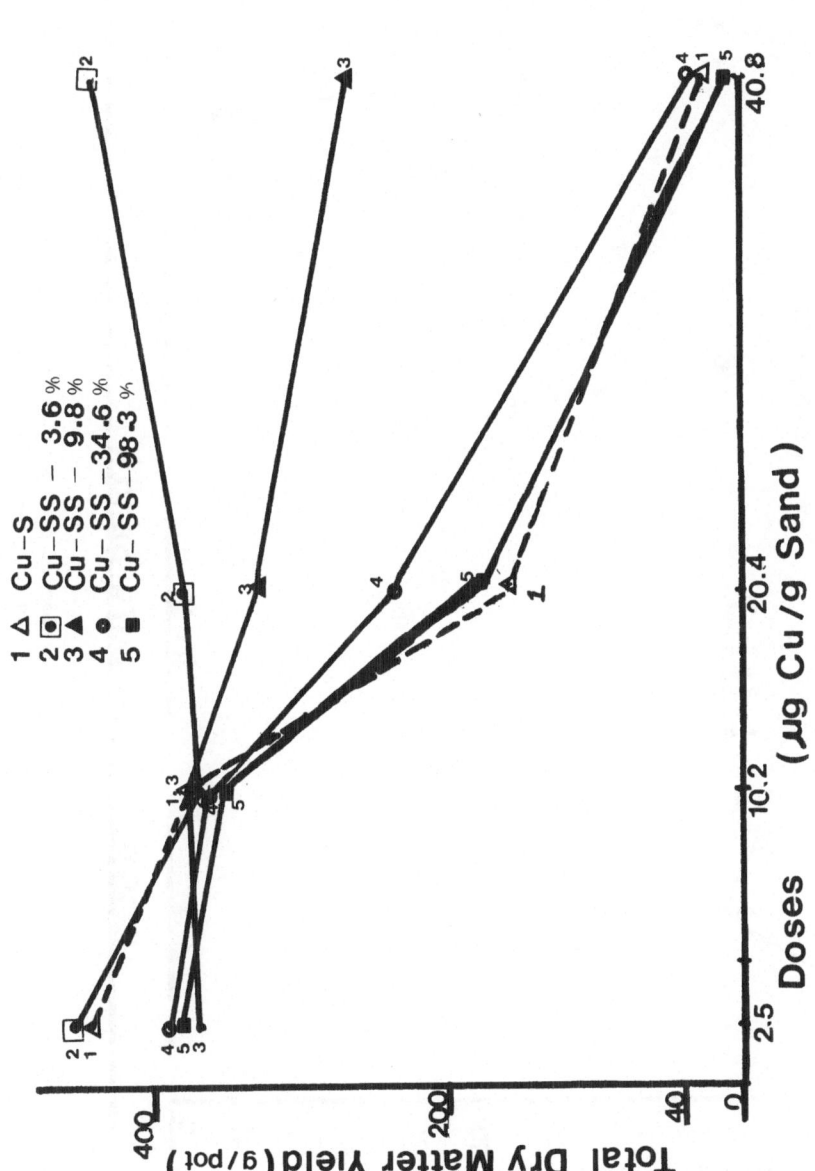

Fig. 1. Effect of graded doses of copper (μg Cu/g sand) and degrees of copper saturation of a sludge sample (Cu-SS-3.6%, Cu-SS-9.8%, Cu-SS-34.6% and Cu-SS-98.3%) on the average dry matter yield of corn (straw + cob + grain + root) in a sand culture.

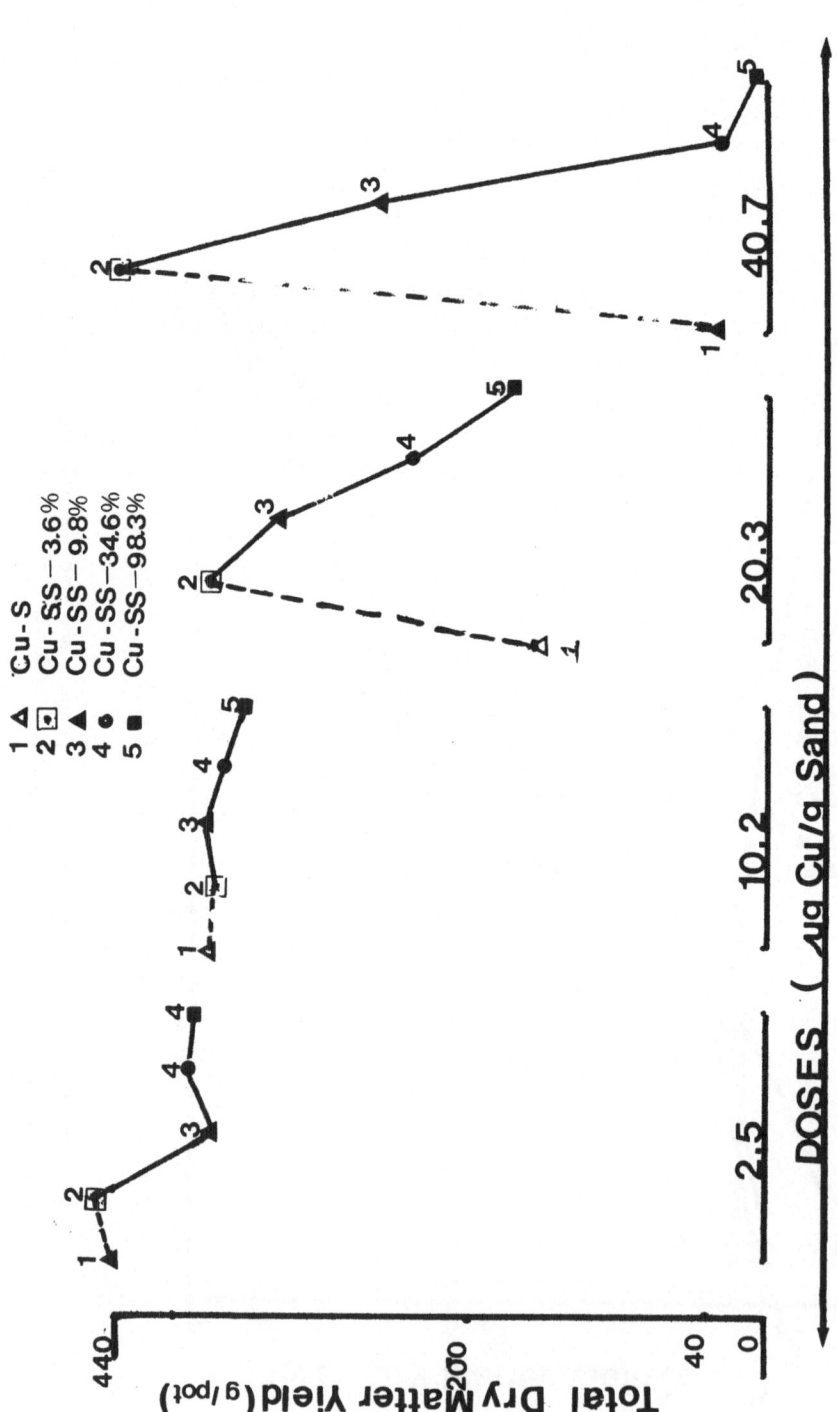

Fig. 2. Effect of degrees of copper saturation of a sludge sample (Cu-SS-3.6%, Cu-SS-9.8%, Cu-SS-34.6% and Cu-SS-98.3%) at four copper doses (μg Cu/g sand) on the dry matter yield of corn (straw + cob + grain + root) in a sand culture.

Cu-Dose: 20.4 μg/g Sand

Control　Cu-S¹　Cu-S　Cu-SS-3.6%　Cu-SS-9.8%　Cu-SS-34.6%　Cu-SS-98.3%

Cu-Dose: 40.8 μg/g Sand

Control　Cu-S¹　Cu-S　Cu-SS-3.6%　Cu-SS-9.8%　Cu-SS-34.6%　Cu-SS-98.3%

Cu-S¹: Copper nitrate — 2.5 μg Cu/g Sand

Cu-S: Copper nitrate — 20.4 μg Cu/g Sand

Fig. 3.　Effect of degrees of copper saturation of a sludge sample (Cu-SS-
3.6%, Cu-SS-9.8%, Cu-SS-34.6% and Cu-SS-98.3%) at two copper doses
(20.4 and 40.8 μg Cu/g sand) on corn growth in a sand culture
experiment.

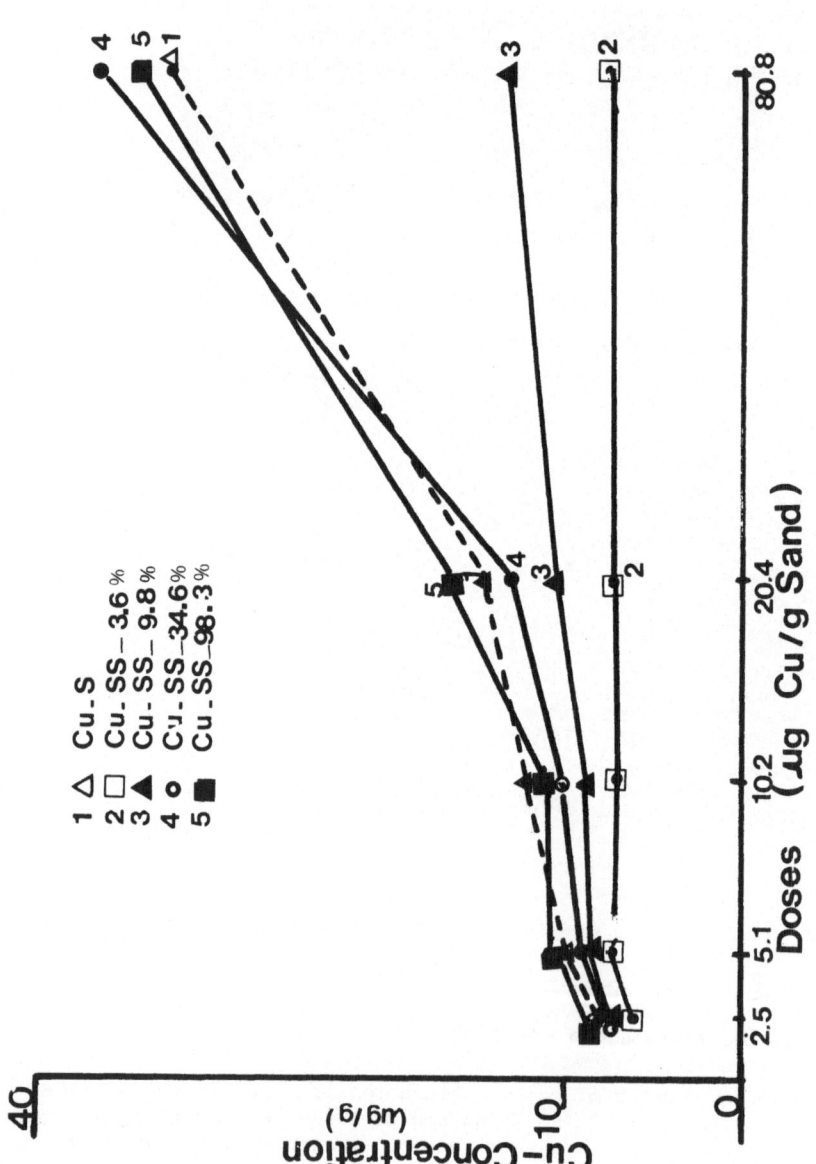

Fig. 4. Effect of graded doses of copper (µg Cu/g sand) and degrees of copper saturation of a sludge sample (Cu-SS-3.6%, Cu-SS-9.8%, Cu-SS-34.6% and Cu-SS-98.3%) on the average copper concentration of corn (straw + cob + grain) in a sand culture.

sample and amount of total copper applied. This observation is substantiated by the fact that there exists a significant relationship between total dry matter yield, degree of copper saturation and copper application levels $(0.668 = r^2)$.

Effect of degree of copper saturation of sludge on the copper concentration of corn

The copper concentration in different parts of corn i.e. straw, cob, grain and root are presented in Table 3. The average copper concentration of corn is graphically presented in Figure 4. The results are as follows:

- The average copper concentration in corn (straw + cob + grain) increased rapidly as the level of application of copper in the form of copper nitrate increased. In the case of a copper application of 2.5 µg/g sand, an average copper concentration of 8.1 ppm was found compared to 32 ppm at an application of 40.8 µg Cu/g sand. The critical level of copper in corn leaves was in the range of 15 - 20 ppm at which level a toxic effect was apparent (Hardy et al., 1967).

- The average copper concentration of corn was found to increase as the sludge sample was progressively saturated with Cu. This observation also holds good for all the levels of application in this experiment.

At the highest Cu dose i.e. 40.8 µg Cu/g sand, the average concentrations of copper in corn were found to be 32, 34, 37 and 13 ppm for different sources used i.e. copper nitrate, Cu-SS-98.3%, Cu-SS-34.6%, Cu-SS-9.8% respectively. In contrast, for the Cu-SS-3.4% treatment, an average copper concentration in corn was found to be 7 ppm.

The data presented in Table 1 show that the sludge copper complexes had a progressively higher amount of Cu as the degree of copper saturation increased. A highly copper saturated complex would have a larger fraction of its total

copper loosely bound as compared to a sludge which is saturated
to a low degree. It is a fact that the metal binding sites of
sludge are heterogeneous and the strength of binding decreases
as the more specific sites are filled. Hence, weakly bound
copper might have adversely affected the corn growth.

The results of this experiment indicate that the
increasing degree of Cu saturation (from 9.8% to 98.3%) of a
sludge sample caused an increasing toxic effect on corn growth,
which is true for all the levels of copper application used in
this experiment. However, the toxic effects were more
pronounced at copper levels of 10.2, 20.4 and 40.8 µg Cu/g
sand. Similar effects were observed on the copper concentration of the corn. In the case of Cu-sludge complex -3.4%
(2 076 µg Cu/g) dry sludge did not affect growth and did not
even enhance the copper concentration of corn; this is true
for all the copper application levels. The limiting value of
copper for sludge in Switzerland and in many other countries
is 1 000 µg Cu/g dry sludge (Furrer et al., 1980; Keller,
1977b).

These results suggest that the effect of sludge bound
metal on the growth of corn should be judged partly by the
degree of copper saturation of sludge and partly by the amount
of total copper applied to the field in sludge.

REFERENCES

Furrer, O.J., 1977. Landwirtschaftliche Verwertung von Klärschlamm;
    Probleme durch Industrieabwässer. Tectilveredlung 12 (6), 244-247.
Furrer, O.J., Keller, P., Haeni, H. and Gupta, S., 1980. Schadstoffgrenz-
    werte - Entstehung und Notwendigkeit. EAS - Seminar 24-26, September
    1980, Basel.
Gupta, S.K., Haeni, H. and Schindler, P., 1980. Mobilisierung und
    Immobilisierung von Metallen durch die organische Bodensubstanz.
    Interner Forschungsbericht, FAC, Liebefeld, Bern.
Hardy, G.W. et al (Eds.), 1967. Soil testing and plant analysis, Part II.
    Soil Science Society of America Special publications Nr 2.
Keller, P., 1976. Zielsetzung und Kriterien bei der Festlegung von Grenz-
    werten für den zulässigen Schwermetallgehalt im Klärschlamm.
    Interner Bericht, FAC, Liebefeld, Bern.
Keller, P., 1977a. Ein Vorschlag zur Kontrolle der landwirtschaftlichen
    Klärschlammverwertung. Landw. Forschung, SH 33/I, 257-273.
Keller, P., 1977b. Zulässige Gehalte des landwirtschaftlich verwerteten
    Klärschlamms an Schwermetallen. Informationstagung: 'Klärschlamm-
    verwertung in der Landwirtschaft'. Schweiz. landw. Technikum,
    Zollikofen.
Kloke, A., 1980. Orientierungsdaten für tolerierbare Gesamtgehalte einiger
    Elemente in Kulturböden. Mitteilungen VDLUFA, Nr. 1-3, 9-11.
Marinsky, J.A., Gupta, S. and Schindler, P., 1980. The Interaction of Cu
    (II) Ion with Humic Acid. (Under Publication).
Schnitzer, M. and Gupta, U.C., 1972. Determination of acidity in soil
    organic matter. Soil Sci. Soc. Am. J. 29, 274-277.

DISCUSSION

<u>Th.M. Lexmond</u> *(Netherlands)*

With regard to the stability of the organic matter from the sewage sludge in the soil, do you expect the total exchangeable acidity to decrease with time?

<u>S. Gupta</u> *(Switzerland)*

Yes, it will certainly decrease with time because the organic matter in the sludge will decompose through micro-biological action. However, at the same time, the organic matter produced by the microbial growth will again complex the copper released through the decomposition.

<u>Th.M. Lexmond</u>

But you may expect some increase in copper activity with time?

<u>S. Gupta</u>

Yes, there will be a temporary increase in activity in the soil solution but later on it will be complexed with dead organic matter. This ties-in with the results Dr. Webber gave this morning on sludge application - one sludge which had about 20 000 ppm copper showed a toxic effect in field conditions, whereas another sludge which had about 2 600 ppm, did not.

<u>P. Worthington</u> *(UK)*

Berrow, in Scotland, did some work on extractable copper over a period of eight years and found no change over that period. I agree that organic matter should break down and release copper, but in practice it is found that this does not occur.

<u>S. Gupta</u>

At our research station our microbiologist found that after one or two years most of the carbon associated with

sludge had been decomposed, thus the copper will be released
in the course of time.

P. Worthington

It will be released, yes, but not necessarily in an
extractable form.

P.H.T. Beckett (UK)

I think Dr. Berrow's figures showed that the half life
of the total organic matter, that is both sludge and new
microbial tissue, was about five years. So, either the heavy
metals are being transferred to other processes, or only a
relatively small amount of organic matter is needed to keep
them immobilised.

S. Gupta

I agree that maybe sludge is decomposed after five years
but there are other exchange sites on the soil components, such
as clay and soil organic matter, which can re-absorb these
heavy metals.

A. Gomez (France)

I think the form of the complexes will be different in
sludge with added copper, compared with natural sludge. In
the first case the complexes will be more soluble than the
larger complexes in the natural sludge.

S. Gupta

Yes, I agree, the complexes can be different. However,
if you add cadmium before processing of the sludge the effect
will be the same as if you add it afterwards because there will
be equilibrium set up in the soil.

A. Gomez

I am not sure about that because French work indicates

a difference between sludge with added metal and sludge which contains the same metal initially. This is because the microbial activity is not the same in the presence or absence of metal, and this results in different kinds of organic matter.

S. Gupta

The addition of 50 000 µg Cu/g sludge to the processing plant will bring microbial activity to a stop.

R.J. Unwin *(UK)*

I would like to address a question to Dr. Gomez. At what stage do you add copper in the preparation of these sludges?

A. Gomez

In the water, with nutrients. The work is being done with a nutrient solution with organic matter on metal; the sludge is made with this solution.

S. Gupta

There will be the same problem at waste water treatment plants where copper comes in different forms depending on the industry of origin.

GENERAL DISCUSSION

R.J. Unwin *(UK)*

May I again pose the question that Dr. Poole raised at
the end of his discussion period, to which he did not get a
response from the audience.   As a soil scientist, I would
like to hear the opinions of the nutritionists on this question
which was, "how long might we expect copper to be fed to pigs
and so present us with the possible problems of disposal of
wastes?".

R. Braude *(UK)*

The answer is simple. Copper is the cheapest available
growth promoter in pig feeding. There are many others but,
economically, this is the most attractive one. Therefore, to
get the maximum economic advantage copper will continue to be
fed until something better is discovered.

J.B. Ludvigsen *(Denmark)*

Or until the general public declines to eat pig liver
any more. Then it will not be so cheap.

R. Braude

But calf liver contains just as much copper as pig liver
from supplemented pigs. There is never any objection to eating
calf liver so therefore there should not be any objection to
eating pig liver from copper supplemented pigs.

J.B. Ludvigsen

The problem is that if the public becomes aware that
copper is used as a growth promoter at the present levels,
they may reject pig liver but not calf liver, even if it is
illogical.

R. Braude

If liver from heavily copper-supplemented pigs, is eaten

a person would have to consume about 20 kg of liver per week
to be in any danger of a copper effect.

J.B. Ludvigsen

But suppose those people are deficient in zinc and
molybdenum!

K.L. Robinson   *(UK)*

I gather that American recommendations now suggest that
the human diet in the USA may be deficient in zinc and that
attention ought to be paid to this situation.

I. Bremner *(UK)*

It has been stated that the US diet is also deficient in
copper.

K.L. Robinson

This is an important point that Dr. Unwin has raised.
It is a point which people do raise, and will continue to
raise, from time to time. After all, the very fact that
this workshop has been convened indicates that there are some
hazardous aspects in relation to feeding copper to pigs. If,
in fact, there are other growth promoters which will do the
same job as copper in performance terms then it can well be
asked, why not use them. However, as Dr. Braude has pointed
out, there would seem to be very strong economic incentives to
use copper and that must be set against the possible hazards.

N.P. Lenis *(Netherlands)*

I don't think the question that has been raised is
relevant at the moment in taking a decision on whether or not
to reduce copper levels.

R.J. Unwin

I take that point. However, I believe Dr. Gasser first
raised the issue earlier today. There is a finite time during

which the present practice can be continued. We can look to
the immediate future and say that there is no problem but that
there are some areas of land in which, if this practice is
going to continue for many years, a problem may build up. If
there is a possibility that there are going to be changes in
the long term, then it would be wrong for us to be considering
penalising the pig producers at the present time - if one could
see that alternatives are going to be available in the longer
term.

N.P. Lenis

Yes, but I think you must make decisions on other grounds
than this.

J.H. Voorburg   (Netherlands)

There is a growth promoter that is much cheaper than
copper and that is to stop castrating male pigs.

R. Braude

That is a red herring I'm afraid. Boars also respond
to copper and you will get improvement by feeding it to boars.
It has nothing to do with castration.

N.P. Lenis

I think that better conditions on the farms will help
much more; better hygienic conditions and so on.

K.L. Robinson

I thought the point about farm hygiene had already been
disposed of. It may be a matter of degree but it has been
pointed out on other occasions that farms on which the hygiene
is satisfactory can still get good responses from feed additives. After all, Dr. Braude's Institute, the NIRD, has
produced many positive results, as have many other research
institutes, and one can hardly question the conditions of
hygiene in such establishments.

D.B.R. Poole

Having started this hare maybe, perhaps I should finish it. It seems to me to be irrational to think that we can legislate for various countries and various conditions on an entirely uniform basis. It also seems irrational to think that in Ireland we might be considering the use of copper sulphate as a soil dressing to raise pasture copper levels and yet be debarred from using copper-rich pig slurry. This is the sort of irrationality that could so easily creep into the situation. However, in saying that, I quite accept that there may be other countries in the Community with an entirely different situation.

SESSION III

EFFECTS ON THE CONSUMER OF MEAT
FROM ANIMALS FED DIETARY COPPER

Chairman: G. Bories

THE EFFECT ON HUMAN COPPER STATUS OF THE CONSUMPTION
OF EDIBLE TISSUES FROM ANIMALS FED Cu-RICH DIETS

G. Bories

INRA, Laboratoire de Recherches sur les Additifs Alimentaires,
180, chemin de Tournefeuille, 31300 Toulouse, France.

ABSTRACT

*The use of large quantities of copper in agriculture for plant pro-
tection or animal treatment (feed additive), gives rise to the problem of
contamination of the food chain, and ultimately to the health of the consumer. This paper attempts to provide answers to the following questions:*

*a)   Does a cumulative process occur in farm animals and, if so, to what
      extent?*

*b)   What is the effect of the consumption of edible tissues from such
      animals on the average or maximum copper contents of human diets?*

*c)   Taking into account physiological requirements and normal dietary
      supplies, does an average increase or occasional high overloading
      represent any hazard for the different consumer groups?*

INTRODUCTION

Hippocrates' prescriptions for the treatment of mental or pulmonary diseases as well as a number of therapeutic uses during the 19th century underline the interest man has always had in copper. But the real biological importance of this element has been established during the last 50 years. An examination of the copious literature, in particular Mason's (1979) excellent review to which we will often refer, shows that positive characteristics, mainly the essential role played by this oligo-element in major enzymatic systems, predominate over adverse effects of cupric ions or organic copper.

The evaluation of copper needs of farm and laboratory animals, as well as of man, has been the main concern of research on copper nutrition until now. At the same time, numerous studies have provided an inventory of plant and animal copper content in order to evaluate man's average daily intake according to his different food habits. During the last decade, new environmental problems have become apparent, such as the effect of long term and large scale utilisation of certain chemicals, some of which accumulate in the food chain. As with other minerals, copper had to be studied; this Workshop Session hopes to bring light on this subject.

Our task was to evaluate the risks for human consumers of an overloading of the copper content of animal products as a result of:

a)   an ingestion of feeds, in particular plants, rich in copper;

b)   the use of copper sulphate as a growth promotant in pigs.

Three main points are examined successively:

a)   the fate of copper in mammals that determine copper transfer in this link of the food chain;

b)   the influence of feed supplementation with copper on copper contents of edible tissues;

c)   the effect of an increase of residual levels in certain
       tissues on human daily intake of copper, in order to
       evaluate the risks for the standard consumer as well as
       consumers who are more vulnerable due to their nutrit-
       ional or physiological state.

COPPER TRANSFER THROUGHOUT THE ANIMAL ORGANISM: MECHANISMS AND
FACTORS THAT DETERMINE ITS EXTENT

     Copper is released from ingested foodstuffs either as
ionic copper, or as a copper-amino acid complex. In the intestinal lumen high molecular weight proteins bind copper and
preferentially release it to the plasma membrane of the luminal
side of the enterocytes. Within the absorption cells a metallo-
thionine rich in sulphydryl groups binds copper through
formation of mercaptide bonds. This cuproprotein serves as a
storage depot and also releases copper to the plasma cell membrane on the serosal side (Figure 1).

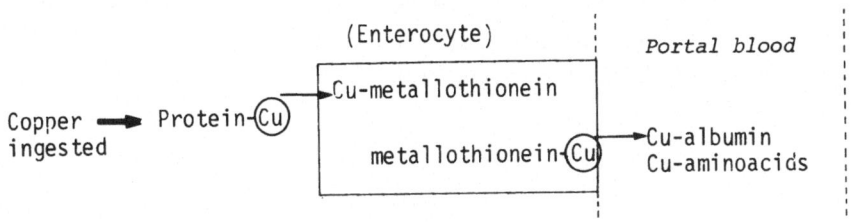

Fig. 1. Copper absorption in mammals

     Major absorption of copper occurs in the stomach and
duodenum: within 1 or 2 hours it appears in serum albumin and
amino acids, then follows a sharp decline as the copper is taken
up by the liver, and subsequently, after 48 to 72 hours, serum
contents increase again as a result of ceruloplasmin synthesis
and blood release; that not immediately extracted by liver
remains attached to serum albumin or amino acids.

     Because copper is both excreted into (mainly through the
bile), and absorbed by, the gastrointestinal tract, faecal
excretion provides a measure of retention but not of true

absorption. Balance studies performed on man have shown that
the mean absorption of an oral dose of 2 mg of copper daily was
56% (range 40 - 70%) (Strickland, 1972). A wide individual
variation must be accepted.

Following release from the intestinal mucosa, copper
becomes bound to albumin and amino acids in the portal blood.
Liver allows a portion of these bound copper complexes to pass
directly to the systemic circulation, where they constitute 7%
of plasma copper. Upon arrival at the liver, copper is released
to hepatocyte cell membrane receptors from which it is transferred to the cytosol where it is bound to metallothionine-like
cuproteins.

In addition to serving as the major pathway of copper
excretion via the biliary tract, evaluated at 0.8 to 1.2 µg per
day in man, the liver releases copper to maintain the labile
pool in the serum and blood cells. However, the major function
of liver is the synthesis of ceruloplasmin. Once synthesised
this cuprotein is released by the liver to the extent that
it comprises about 93% of plasma copper (Figure 2).

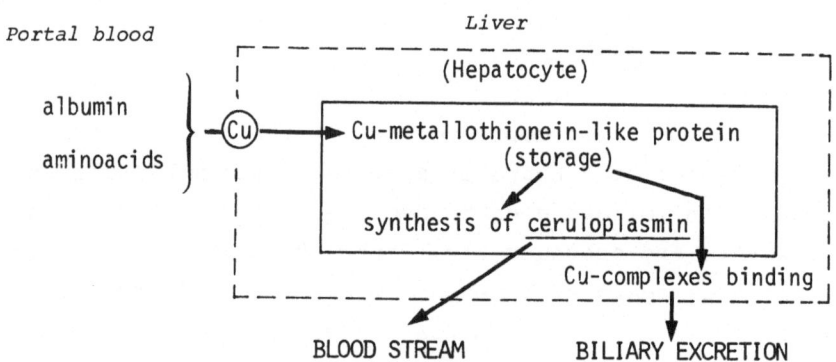

Fig. 2. Copper transport and disposition in mammals

The following copper balance in man may be retained (Cartwright and Wintrobe, 1964): from a daily intake of 2.0 to 5.0 mg, 0.6 to 1.6 mg (32%) is absorbed; 0.01 to 0.06 mg is excreted in urine, 0.1 to 0.3 mg passes directly into the bowel, and 0.5 to 1.3 mg is excreted in bile (Figure 3).

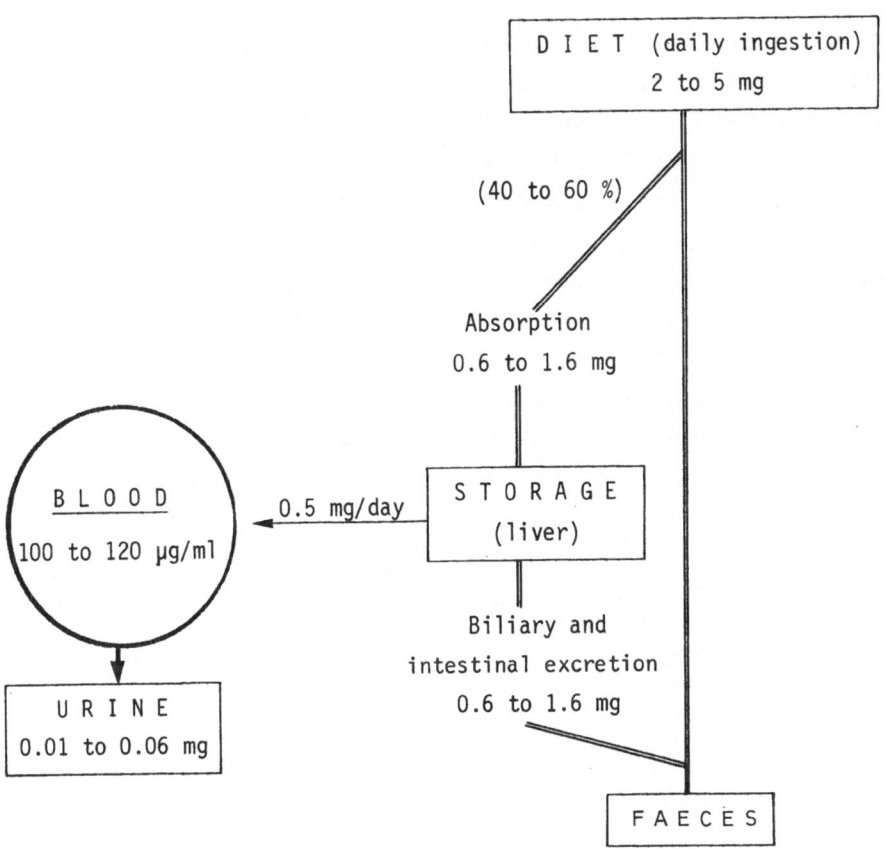

Fig. 3. Metabolic balance of copper in man.

The result is that almost 0.5 mg of the absorbed copper is released into the bloodstream, which corresponds to the real needs of the organism. Excess intake means an increase of biliary excretion, followed by copper accumulation when this pathway becomes saturated.

Another point to be remembered is that in the organism copper only occurs bound to proteins or amino acids, never in ionic form.

EFFECT OF COPPER SUPPLEMENTATION OF FEEDS ON COPPER STATUS OF FARM ANIMALS

This analysis is mainly based on the results compiled by Meyer and Kröger (1973) from numerous trials carried out on pigs. The main conclusions are as follows:

1. Copper is not distributed evenly between the different tissues of pigs fed a standard diet without copper addition. By far the most copper accumulation occurs in the liver compared with other organs, the decreasing order of concentrations in the other tissues or organs being: kidneys $\gg$ spleen > lungs > muscle > fat.

2. Feed supplementation with 250 ppm copper sulphate, a common dosage level, results in an increase of the copper content of most tissues (Table 1) and especially of the liver.

TABLE 1

COPPER CONTENT OF DIFFERENT TISSUES AND ORGANS OF PIG (90-100 kg) FED A DIET CONTAINING 250 ppm Cu SO_4

Organ or tissue	μg per g wet tissue	% of control value
Liver	250	2 500
Kidneys	10 to 13	160
Muscle	1.3	130
Fat	-	-

From H. Meyer and H. Kröger

3. Copper concentration in the liver increases with feed dosage (Table 2) and with length of time of administration (Table 3). A more recent study (Nadazin et al., 1977) carried out on pigs showed that 50 and 250 ppm copper supplementation resulted in

liver residues of 8.9 and 88.1 ppm respectively. The
cumulative process is not linear, storage being proportionally
much higher with increasing intake of copper, especially over
100 ppm in the diet (Table 2); this phenomenon is directly
related to the saturation of the metabolic pathway, as has
already been mentioned.

TABLE 2

EFFECT OF COPPER SULPHATE ADDITION TO PIG DIET ON Cu CONTENT OF THE LIVER
AT SLAUGHTER (90-100 kg)

Copper sulphate (ppm)	Cu concentration in liver (µg per g wet tissue)			Number of	
	lower	(average)	higher	trials	animals
0	1.6	**12**	57	22	261
60	12.6	**17**	29	2	18
125	7.3	**57**	273	7	87
250	8	**256**	890	27	443
500	-	**817**	1 556	4	28

From H. Meyer and H. Kröger

TABLE 3

INCIDENCE OF LENGTH OF TIME OF ADMINISTRATION OF COPPER SULPHATE (250 ppm)
ON Cu CONTENT OF PIG LIVER AT SLAUGHTER

Duration of Supplementation	Cu concentration (µg per g wet tissue)		
	15-95 kg	15-56 kg	15-84 kg then withdrawal→ 95 kg (2 weeks)
Trial 1	168	34	57
Trial 2	93	17	-
Trial 3	98	98	-

From H. Meyer and H. Kröger

4. Copper concentration in the liver decreases following with-drawal of the Cu source from the diet (Table 3). It also decreases after birth when copper stored in the foetal liver is gradually mobilised to compensate for the very low content in milk (Table 4).

TABLE 4

EVOLUTION OF COPPER CONCENTRATION IN PIGLET LIVER

	Cu content µg per g of wet tissue
New born piglet	77
Up to 15 kg weight without Cu administration	37
After 150 ppm Cu SO_4 administration (15 to 25 kg weight)	79
Up to 60 kg weight without Cu administration	12
After 150 ppm Cu SO_4 administration (50 to 60 kg weight)	129

From H. Meyer and H. Kröger

5. The chemical form of copper determines the extent of absorption (Table 5). It has been proved that less ionic copper, such as copper sulphate, is absorbed than copper in the form of organic complexes with proteins or amino acids (Kirchgessner and Grassmann, 1970).

Another conclusion that can be drawn is the great vari-ability of individual responses, as illustrated by Table 1. The interaction of other elements such as zinc, iron, molybdenum or sulphur on copper metabolism is now well established (Miller and Engel, 1960; Pitt, 1976). Even if monogastrics are less sensitive than ruminants to unbalanced microelement intakes, ratios such as copper versus molybdenum have a big effect on copper absorption, disposition and storage. Modern diets assess the best ratios, and it may be expected that the highest storage values encountered 10 to 15 years ago, as reported in Meyer and Kröger (1973), would now be sharply decreased.

TABLE 5

EFFECT OF COPPER CHEMICAL FORM ON Cu-RETENTION IN PIG LIVER AT SLAUGHTER (90-110 kg)

Chemical form	ppm of Cu	Cu concentration in liver (µg per g wet tissue)
Cu S	250	17
Cu O	250	23
Cu CO$_3$	250	167
Cu SO$_4$	250	256
Cu SO$_4$	100	12
Cu-methionate	100	12.3

Experiments carried out on calves (Scholz, 1976) indicate that a 250 fold increase of liver copper content (2 849 ppm) results from the addition of 150 ppm copper sulphate in the diet when compared with control animals whose diet only contained 30 ppm 'total' copper (11.3 ppm). The level was only doubled in the kidneys and the concentration in the muscles was unchanged. As for pigs, copper storage in the liver decreases 5 fold after withdrawal of copper sulphate from the diet for a week.

EFFECT OF CONSUMPTION OF COPPER-RICH ANIMAL PRODUCTS ON HUMAN COPPER STATUS. TENTATIVE EVALUATION OF HAZARDS TO THE CONSUMER.

Exhaustive studies of many copper containing proteins distributed throughout different tissues and blood, and their role in the metabolic process, have underlined the biological importance of copper for mammals. Copper requirements for man have been evaluated from balance experiments carried out on infants and adults. Dietary allowances (optimal levels) have been proposed by the WHO (1973 report) and the NRC (1974 statement). The requirements of infants and children have been estimated at between 0.05 to 0.1 mg/kg body weight/day. A wide acceptance of 2.0 to 2.5 mg for adults has been retained, but more recent studies indicate that levels less than 2 mg, and sometimes not much more than 1 mg, may maintain a positive copper balance.

Copper is supplied through food and drinking water. Copper is ubiquitous in plants and animals, and it is well recognised that its level varies greatly in foods. A recent and complete compilation of Pennington and Calloway (1974) establishes a list by decreasing order of concentrations: the richest sources are liver, crustaceans and shellfish (20 to 50 mg/kg), followed by nuts and seeds, high-protein cereals, dried fruit, poultry, fish, meat, vegetables; cows' milk contains only 0.05 to 0.2 mg/l. Average daily intake of copper throughout most western-style mixed diets has been evaluated to 2 to 4 mg (Underwood,1977). Varying food habits mean there is a wide range of values: 1.5 mg for inhabitants of some Polynesian islands, to 5.8 mg for populations in India consuming rice and wheat diets.

Moreover, day to day copper intake reaches extreme values: a survey of lunches served to 6th grade children in USA schools (Murphy et al., 1971) indicates an average content of 0.34 mg/day, but a range of 0.06 to 2.19 mg. It is difficult to propose a diet where the copper content is less than 0.5 mg/day (Schroeder et al., 1966), however it is easy to compose copper rich meals (30 to 40 mg per day), using oysters, liver and high protein cereals. Baby foods containing beef liver and cereals may reach high copper values such as 2.64 and 1.85 mg/100 g (Hughes et al., 1960).

The effect of the consumption of copper enriched animal products on the normal dietary supply offers different aspects. It has already been established that a 250 ppm supplementation of pig diets increases the residual copper contents not only of

liver but also, to a very limited extent, that of most tissues
and organs. On the basis of the quantitative contribution of
meat to the whole diet, doubling the muscle copper content results in an increase of only about 30% of the average daily
intake. Liver copper contents of animals fed standard diets
without the addition of copper, are about 10 times higher than
in other tissues, even more where ruminants are concerned; however these data have already been taken into account when cal-
culating average copper intake. In the worst situation it has
been established that the same 250 ppm copper supplementation
may increase the copper content in pig liver by 50 times. In
spite of this dramatic increase, if liver consumption may be
estimated at about 1 - 2% of annual meat intake, it can be
calculated that the entire quantity of copper supplied by meat
over this period is only doubled. Although on an average basis
the consumption of high copper content tissues contributes
markedly to the increase of copper intake, it still stays in
the range of the normal variability due to food contents and
food habits. However pig liver consumption is not distributed
evenly throughout the year, but is limited to occasional intake. Therefore, although the average intake increase does not
present major problems, our attention must be focused on the
consequences of transient overloading, for example 20 to 40 mg
eaten in a single day.

Much of the information on copper toxicity comes from
reports of accidental or intentional poisoning. Data concerning
oral intake necessary to produce symptoms of toxicity are very
rare. It has been mentioned that the ingestion of 10 to 15 mg
of inorganic copper causes nausea, vomiting and diarrhoea, and
larger doses cause haemolysis. Most toxic incidents recorded
in more normal circumstances are related to water contamination
by pipe corrosion and defective copper-lined containers. The
symptomatology of copper toxicity has been described precisely
from a copper metabolic abnormality occurring in man which is
called Wilson's disease. This defect in ceruloplasmin synthesis
results in copper accumulation in liver, brain and kidneys with
irreversible damage. To our knowledge similar clinical signs

have never been correlated with high copper dietary intake. This means that adjustment of dietary supplies with metabolic requirements is not a crucial point. Above 1 to 2 mg per day, a minimal intake nearly always supplied, copper metabolic characteristics such as slow turnover rate, strong homeostatic mechanisms, and considerable biliary excretion capabilities, determine an extended ability to 'buffer' occasional and even transient copper overloading. Thus it has been mentioned previously that levels of 30 mg copper per day may easily be reached when eating oysters or crustaceans: in these rare situations no adverse effects have been described. Moreover, no direct correlation can be established with the toxicological evaluation of inorganic copper that led to the FAO/WHO statement of 10 to 15 mg per day as a 'dose generally recognised as safe', ionic copper being generally considered as more toxic. Indeed the compared biodisponibility of copper in protein or amino acid complexes and ionic form has not been investigated until now.

However attention must be drawn to particular situations, e.g. children fed baby foods. A more limited variety of available foods, therefore probably a more frequent intake of liver rich preparations, could considerably increase the whole copper intake. From our point of view pig liver from animals fed copper supplemented diets should be avoided for such use.

CONCLUSION

The survey of the effect of copper rich animal products on human copper status is reassuring. Even though high concentrations may be reached in pig liver, involving occasional overloading of consumers' intake, mammals have a powerful homeostatic mechanism based on biliary excretion and temporary storage of copper. The result is that copper intoxication has never been described with 'organic-copper' carried by food. Only baby foods containing liver should be controlled in order to avoid an excessive copper content.

REFERENCES

Cartwright, G.E. and Wintrobe, M.M., 1964. Copper metabolism in normal
  subjects Am. J. Clin. Nutr. 14, 224-232.

Hughes, G., Kelley, V.J. and Stewart, R.A., 1960. The copper content of
  infant foods. Pediatrics, 25, 477-484.

Kirchgessner, M. and Grassmann, E., 1970. The dynamics of copper absorption. Proceedings Intern. Symposium Trace element metabolism in
  Animals p. 277 (Mills, C.F. ed.). Livingstone, Edinburgh and London.

Mason, K.E., 1979. A conspectus of research on copper metabolism and
  requirements of man. J. Nutr. 109,1979-2066.

Meyer, H. and Kröger, H., 1973. Kupferfütterung bein Schwein. Übers.
  Tierernahrg. 1, 9-44.

Miller, R.F. and Engel, R.W., 1960. Interrelations of copper, molybdenum
  and sulfate sulfur in nutrition. Federation Proc. 19, 666-677.

Murphy, E.W., Page, L. and Watt, K., 1971. Trace minerals in type A school
  lunches. J. Am. Diet. Assn. 58, 115-122.

Nadazín, M., Dzinić, M., Papić, D. and Bukojević, J., 1977. Interdependence
  of copper concentration in the feed and liver parenchyma of pigs.
  Veterinaria, Yugoslavia, 26, 49-58.

Pennington, J.T. and Calloway, D.H., 1974. Copper contents of foods. J. Am.
  Diet. Assn. 63, 143-153.

Pitt, R., 1976. Molybdenum toward interactions between copper, molybdenum
  and sulphate. Agents Actions, 6, 758-767.

Scholz, H., 1976. Cu content of organs of veal calves during and after
  manifest excess of dietary copper. Deutsche Tierarztliche Wochenschrift
  83, 61-62.

Schroeder, H.A., Nason, A.P., Tipton, I.H. and Balassa, J.J., 1966.
  Essential trace metals in Man : copper. J. Chron. Dis. 19, 1007-1034.

Strickland, G.T., Beckner, W.M. and Leu, M.L., 1972. Absorption of copper
  in homozygotes and heterozygotes for Wilson's disease and controls:
  Isotope tracer studies with 67Cu and 64Cu. Clin. Sci. 43,617-625.

Underwood, E.J., 1977. Trace elements in Human and Animal Nutrition, ed.
  Academic Press, New York.

SESSION IV

INDUSTRIAL PROBLEMS

Chairman: J.K.R. Gasser

COPPER AS A GROWTH PROMOTER

B.C. Cooke

Dalgety Spillers Limited,
The Promenade, Clifton, Bristol, UK.

INTRODUCTION

The subject of copper as a growth promoter for growing pigs was first raised by Braude et al. in the early 1950s. As a result of observations of pigs that were licking copper pipes within their pens, they instigated experimental work to ascertain whether pigs would respond to higher levels of copper in their diet. Following this work, they produced the recommendation that the addition of 250 mg/kg of copper to the diets of growing pigs would lead to improved growth rates and feed conversion efficiency. As a result of these publications by Braude et al., it became common practice in the UK and other countries in the world to add about 2 lb or 1 kg of copper sulphate to each tonne of feed for growing pigs.

However, when the EEC was set up, none of the original six countries involved commonly used high levels of copper. Thus, the legislation was written with a maximum of 125 mg/kg of total copper in the diets of growing pigs.

In the UK, whilst the Agricultural Act laid down this maximum, CAFMA, the Compounded Animal Feedstuffs Manufacturers Association, applied for a product licence under the Medicines Act to enable the continued use of 200 mg/kg. This product licence was granted, and has since been transferred to UKASTA. Application was made on our joining the EEC for the inclusion of copper at 200 mg/kg in Annex 2 of Directive 70/524.

However, it is laid down that no product should stay in Annex 2 on a permanent basis, the aim being either to transfer it to Annex 1, or to prevent its use completely within a period of three to five years. Thus, the temporary arrangement

of placing copper for growing pig feeds in Annex 2 enabled
the UK and any other member country which so desired to allow
its use within the boundaries of that state.

About three to four years ago it became obvious that
further data would be required by various committees and legis-
lators within the EEC if the pig industry was ever to obtain
permanent permission for the use of copper at 200 mg/kg in
growing pig feeds. Originally the entry of copper in Annex 2
expired at the end of 1977. Early in that year it was decided
that, as product licence holders, UKASTA should undertake a
survey of all the published literature on the subject of copper
in pig feeds, with a view to justifying the continued use of
200 mg/kg.

All the published literature on the subject of copper
supplementation of growing pig diets between 1955 and 1975 was
studied. 159 references were found for consideration. Many
of these references referred to more than one experiment, and
a number to more than one stage of growth during the overall
trial. Some of the references found were unsuitable due to
the fact that the data were incomplete, or that they merely
duplicated other references, so that finally, 1 594 results
were entered onto the computer, and various regression equations were obtained as a measure of the response to copper
supplementation. The mean performance data from these results
are given in Table 1.

TABLE 1

OVERALL MEAN PERFORMANCE DATA

Variable	Mean	Standard deviation
Liveweight gain kg/day	0.627	0.145
Feed consumption kg/day	1.943	0.725
Feed conversion ratio	3.124	0.661
Copper level mg/kg	230	

Statistical analysis was carried out in order to ascertain which type of curve best described the response to copper, and this showed that the multiple regression explained the highest proportion of the variability between experimental animals and controls. This type of curve was thus decided upon as the one which would be used to study the various factors which might affect the response of growing pigs to the addition of copper to their diets.

The curves shown in Figure 1 relate to the response in growth rate to the addition of copper. All the equations illustrated by these lines were highly significant. Investigation of the responses in feed conversion efficiency are illustrated in Figure 2, where it can be seen that the curves are almost identical mirror images of the growth curves.

However, due to greater variability in the data, these feed conversion curves do not reach the same levels of significance as do the growth curves. They clearly indicate, however, that the extra growth obtained from the addition of copper is realised without the intake of a comparable amount of extra feed.

Having obtained these overall effects, the study then went on to investigate various factors which might modify the response. These factors are growth rate of the control, weight range of the pig, source of cereal and protein, feeding system and country of origin. The effect of these factors on the growth rate response are discussed in the remainder of this paper.

GROWTH RATE OF CONTROL

It was found from this part of the study that the growth rate of the control animals had a significant effect upon the response to copper. Faster growing pigs responded less. This is illustrated by the curves in Figure 3. This is a fairly normal phenomena in studies of growth promoters. All causes of

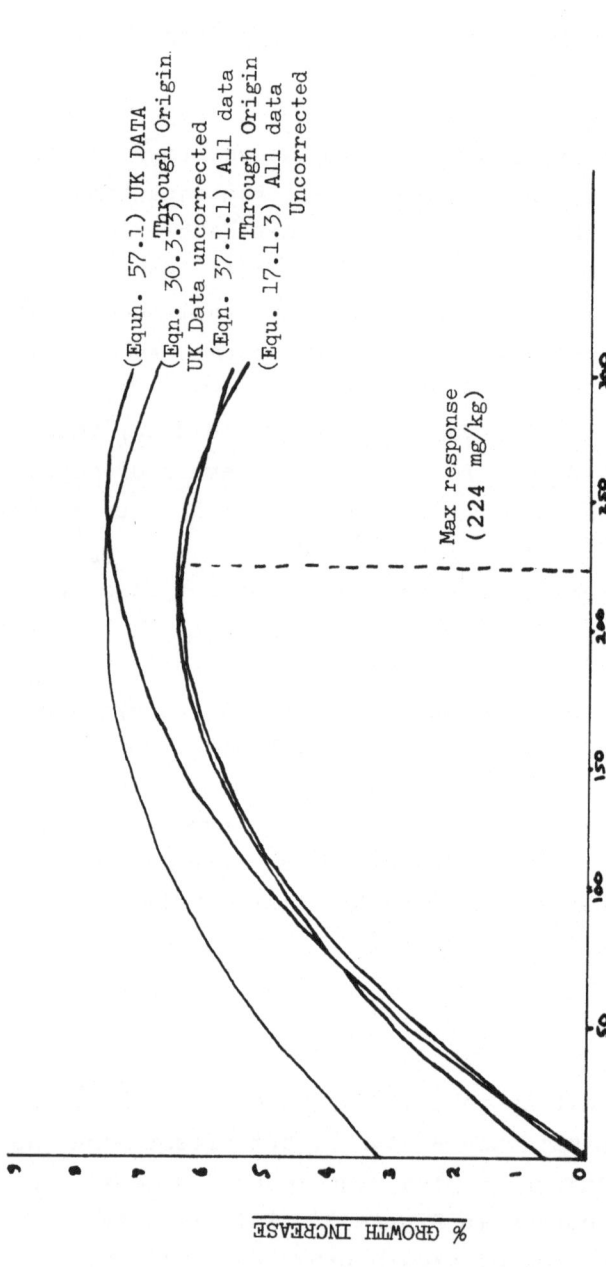

Fig. 1. Growth response of pigs to copper in diets evaluated for LWGC = 0.6 kg/day

Equn. 57.1 UK through origin. Live wt gain effect = (0.373×10^{-3}) Cu − (0.752×10^{-6}) Cu^2 − $(0.197 \times$ (live wt gain of control − mean))

Equn. 30.3.3. UK uncorrected. Live wt gain effect = $0.136 + (0.259 \times 10^{-3})$ Cu − (0.611×10^{-6}) Cu^2 − $(0.191 \times$ live wt gain of control)

Equn. 37.1.1. All through origin. Live wt gain effect = (0.349×10^{-3}) Cu − (0.780×10^{-6}) Cu^2 − $(0.151 \times$ (live wt gain of control − mean))

Equn. 17.1.3 All uncorrected. Live wt gain effect = $0.0944 + (0.324 \times 10^{-3})$ Cu − (0.751×10^{-6}) Cu^2 − $(0.151 \times$ live wt gain of control)

ADDED COPPER IN FEED (mg/kg)

% GROWTH INCREASE

Max response (224 mg/kg)

(Equn. 57.1) UK DATA

(Eqn. 30.3.3) Through Origin UK Data uncorrected

(Eqn. 37.1.1) All data Through Origin

(Equ. 17.1.3) All data Uncorrected

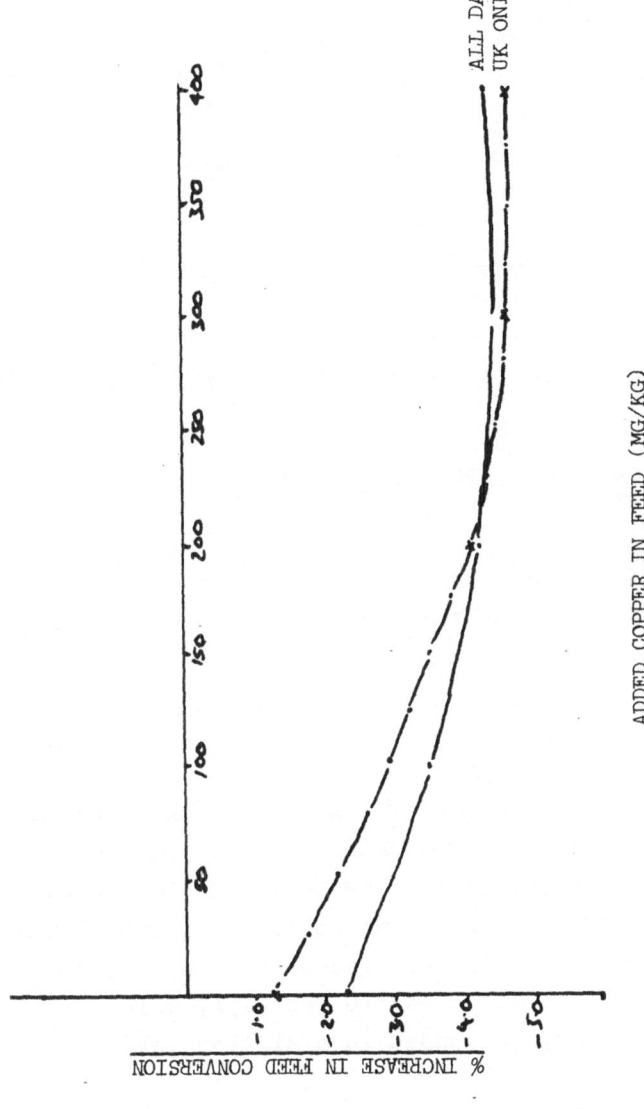

ADDED COPPER IN FEED (MG/KG)

Equn 17.2.3 All uncorrected
Feed conversion effect = $0.1128 - (0.430 \times 10^{-3})$ Cu + (0.707×10^{-6}) Cu^2 –
$(0.0594 \times$ f.c.e. of control)

Equn 25.6.3 UK uncorrected
Feed conversion effect = $0.0475 - (0.618 \times 10^{-3})$ Cu + (0.895×10^{-6}) Cu^2 –
$(0.0279 \times$ f.c.e. of control)

Fig. 2. Response in feed conversion efficiency to copper in the diet. Evaluated for FCRC = 3.124.

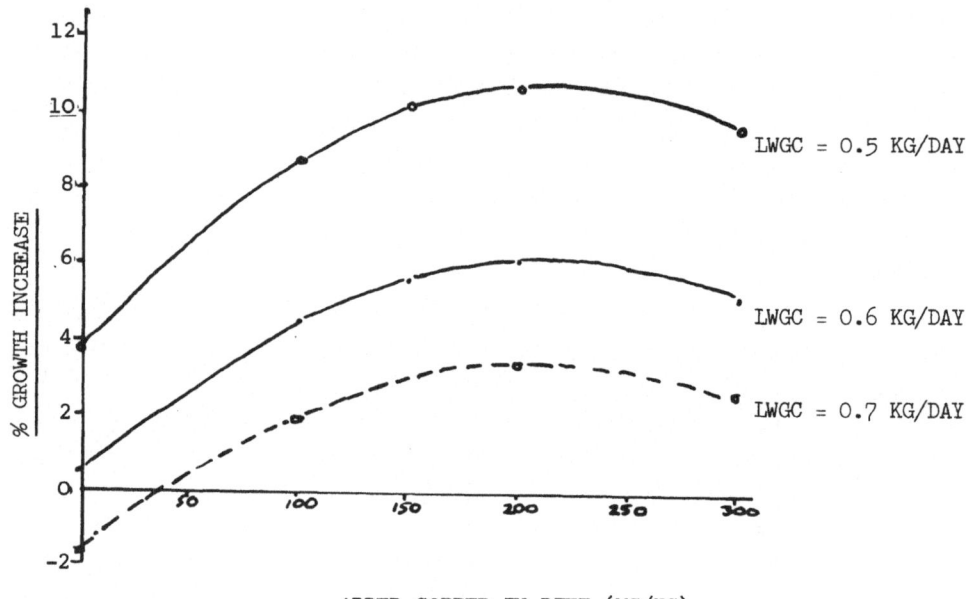

ADDED COPPER IN DIET (MG/KG)

Equn 17.1.3 All uncorrected
Live wt gain effect = 0.0944 + (0.324 x 10^{-3}) Cu - (0.751 x 10^{-6}) Cu^2 (0.151 x
live wt gain of control)

Fig. 3.   Growth response of pigs to copper in the diet:-
          For different growth rates

slower growth rate (stress, disease, poor pigs, poor environment, etc.) tend to lead to a greater response to the growth
promoter.

FEEDING SYSTEMS

        One factor which might affect growth rate is the feeding
system. Figure 4 illustrates curves obtained from pigs which
were fed semi-ad-lib and restricted. The restricted fed pigs
gave a 9% response in growth rate to the addition of 308 mg/kg
of copper, whilst the semi-ad-lib fed pigs only gave a 4.3%
response to 193 mg/kg of copper.   Ad-lib fed pigs, which were
fed to appetite twice a day, responded in a similar fashion to
the semi-ad-lib fed pig.

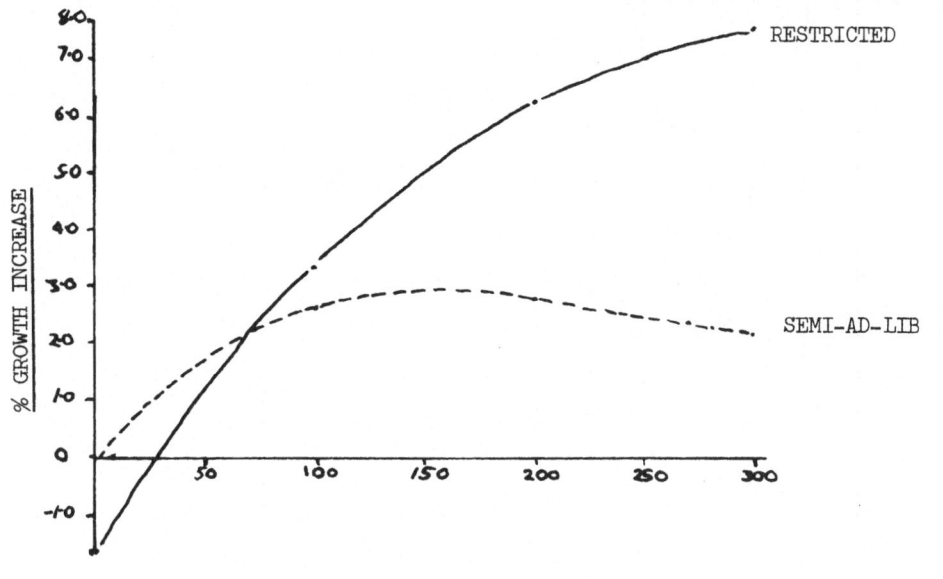

Equn 18.3.3 Semi-Ad-Lib
Live wt gain effect $= 0.145 + (0.268 \times 10^{-3})$ Cu $- (0.693 \times 10^{-6})$ Cu^2 $-$
$(0.249 \times$ live wt gain of control)

Equn 18.5.3. Restricted
Live wt gain effect $= 0.0271 + (0.350 \times 10^{-3})$ Cu$- (0.569 \times 10^{-6})$ Cu^2 $-$
$(0.0614 \times$ live wt gain of control)

Fig. 4. Growth response of pigs to copper in diets on different feeding
         regimes.

AGE/WEIGHT OF PIGS

        The average curve referred to earlier clearly indicates
a reduced response to levels of copper above 250 mg/kg. There
could be many reasons for this response, one of which may be
total copper intake. It could be that pigs which are on copper
for a long period take in an amount above a certain threshold,
and above this figure, the response falls away. One way of
investigating this point is to look at the response of pigs at
different ages. However, this proved almost impossible to do
due to the lack of data on age of pigs involved in the trials.
So, as an alternative, the effects at different weight ranges
were studied. The results of this study are shown in Figure 5.

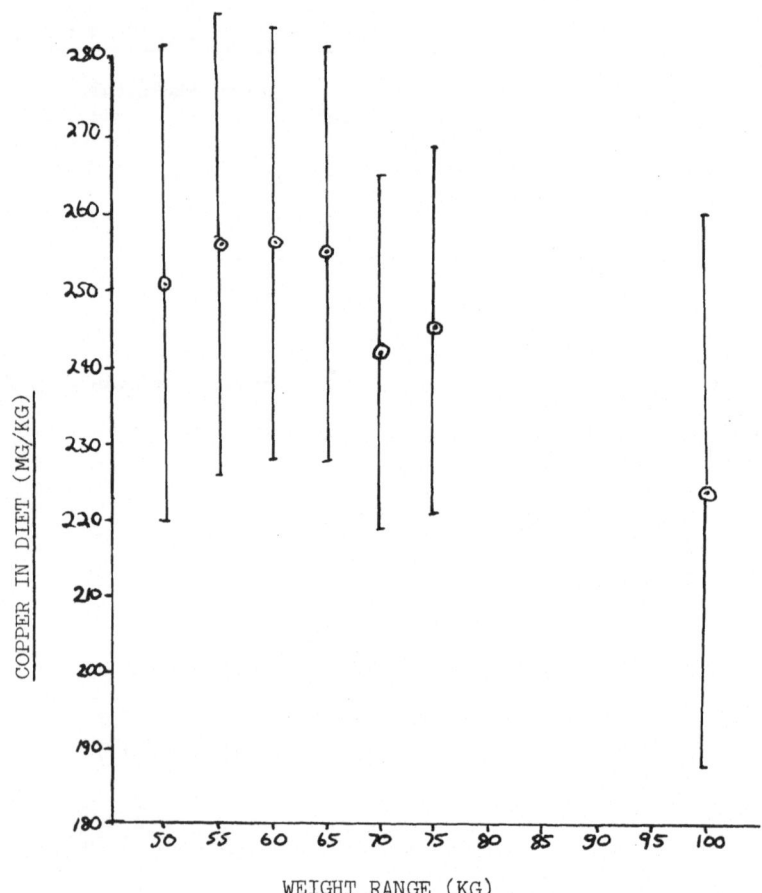

Equn. 45.1 Live wt 4 - 50 kg. Live wt gain effect = (0.288×10^{-3}) Cu - (0.574×10^{-6}) Cu^2 - $(0.170 \times$ (live wt gain of control-mean))

Equn. 45.3 Live wt 4 - 55 kg. Live wt gain effect = (0.322×10^{-3}) Cu - (0.630×10^{-6}) Cu^2 - $(0.180 \times$ (live wt gain of control-mean))

Equn. 45.5 Live wt 4 - 60 kg. Live wt gain effect = (0.337×10^{-3}) Cu - (0.655×10^{-6}) Cu^2 - $(0.173 \times$ (live wt gain of control-mean))

Equn. 45.7 Live wt 4 - 65 kg. Live wt gain effect = (0.335×10^{-3}) Cu - (0.657×10^{-6}) Cu^2 - $(0.164 \times$ (live wt gain of control-mean))

Equn. 45.9 Live wt 4 - 70 kg. Live wt gain effect = (0.339×10^{-3}) Cu - (0.701×10^{-6}) Cu^2 - $(0.187 \times$ (live wt gain of control-mean))

Equn. 45.11 Live wt 4 - 75 kg. Live wt gain effect = (0.359×10^{-3}) Cu - (0.733×10^{-6}) Cu^2 - $(0.188 \times$ (live wt gain of control-mean))

Equn. 37.1.1 Live wt 4 - 100 kg. Live wt gain effect = (0.349×10^{-3}) Cu - (0.780×10^{-6}) Cu^2 - $(0.151 \times$ (live wt gain of control-mean))

Fig. 5. Growth response of pigs to copper in the diet for various liveweight ranges

These data indicate that as pigs get older, the optimal level of copper in the diet falls from about 250 mg/kg in the weight range of 4 - 50 kg liveweight, to 224 mg/kg in the range of 4 - 100 kg liveweight. There is some indication of a reduction in the level of response in the older pigs, but this is not as dramatic as the reduction in the optimal level of copper. However, these data clearly indicate that for the older pig of over 70 - 75 kg liveweight, for instance, lower levels of copper could probably be used without any marked adverse effect upon growth rates and feed conversion efficiency. This indeed has been the recommendation in Germany, where the maximum age of 16 weeks has been put on the use of 200 mg/kg, the level thereafter being 125 mg/kg.

It must, however, be made clear that the imposition of a reduction in copper at any weight less than 75 kg liveweight, or approximately 20 - 22 weeks of age, would be virtually impossible to administer on a farm. It would mean that feed compounders would have to produce diets with the same nutrient content, but with different copper levels. With different ages and weights of pig within one house, which are being fed from one bulk system, many farmers who are feeding bulk compounds would have to put up a second bulk bin or feeding system which might prove economically prohibitive.

SOURCE OF CEREAL AND PROTEIN

The next factor which was considered was whether the response to copper would be different if there were vast differences in the basal diet, for obviously within the EEC, the diets vary tremendously in both the source of cereals and proteins. In the UK and Ireland, the cereal is usually barley, whereas in southern European countries, much more maize and hardly any barley may be used, whilst on the protein side, the UK and Eire have tended traditionally to use fish meal, but in other parts of Europe and over recent years in the UK and Eire, soya has been the main source of protein.

Figure 6 shows curves obtained from data where the predominant cereals were barley and maize and the predominant proteins were fish meal and soya. These curves clearly indicate that the source of energy and protein has little effect upon the level of response which is obtained from the addition of copper.

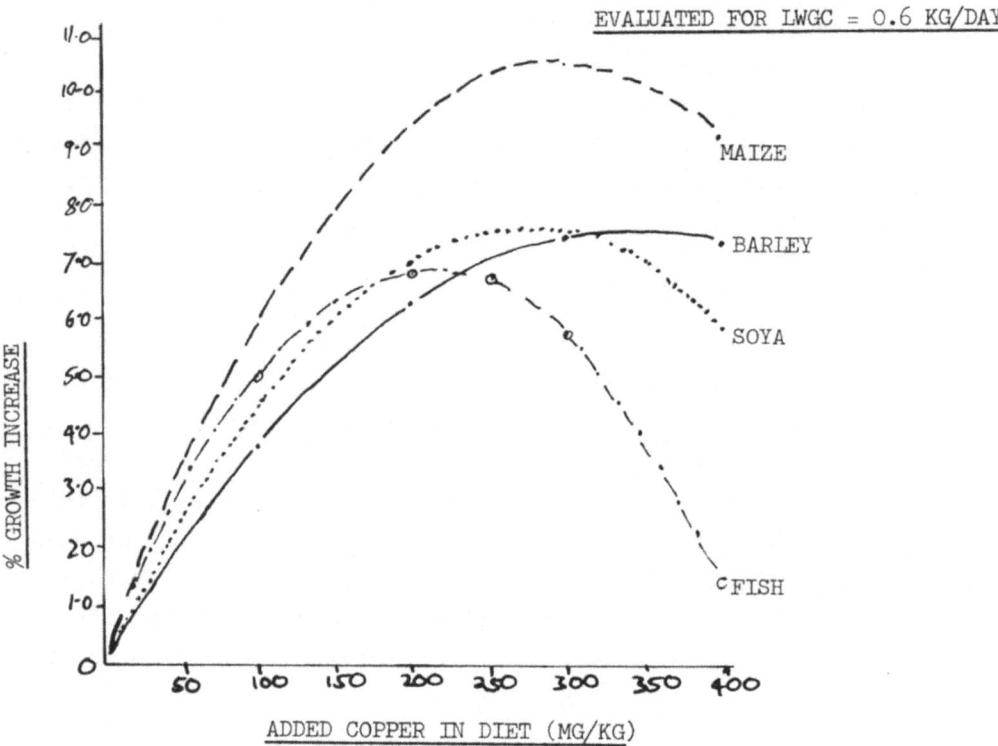

EVALUATED FOR LWGC = 0.6 KG/DAY

ADDED COPPER IN DIET (MG/KG)

Equn. 38.1.1 Barley
Live wt gain effect = (0.269×10^{-3}) Cu $- (0.4 \times 10^{-6})$ Cu^2 $- (0.132 \times$ (live wt gain of control $-$ mean))

Equn. 38.4.1 Maize
Live wt gain effect = (0.424×10^{-3}) Cu $- (0.716 \times 10^{-6})$ Cu^2 $- (0.124 \times$ (live wt gain of control $-$ mean))

Equn. 38.7.3 Soya
Live wt gain effect = (0.334×10^{-3}) Cu $- (0.615 \times 10^{-6})$ Cu^2 $- (0.152 \times$ (live wt gain of control $-$ mean))

Equn. 38.10.3 Fish
Live wt gain effect = (0.397×10^{-3}) Cu $- (0.942 \times 10^{-6})$ Cu^2 $- (0.0165 \times$ (live wt gain of control $-$ mean))

Fig. 6.   Growth response of pigs to copper in the diet - for different basal diets.

COUNTRY OF ORIGIN

When the initial data for this survey were studied, it appeared that the American data were somewhat different to those from other countries. It was therefore decided that analysis should be carried out on a country basis where possible. Only in the cases of the UK, Canada and the USA were there enough references to enable a sensible statistical study to be carried out. The results of this study are shown in Figure 7.

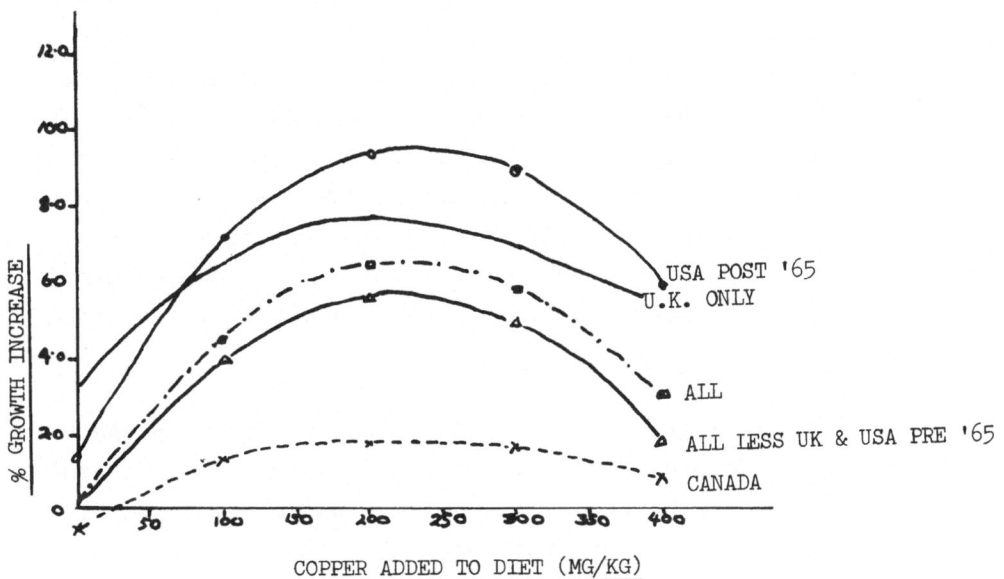

Equn. 46.5 USA Post '65. Live wt gain effect = 0.203 + (0.369 x 10^{-3}) Cu - (0.784 x 10^{-6}) Cu^2 - (0.316 x live wt gain of control)

Equn. 30.3.3 UK only. Live wt gain effect = 0.136 + (0.259 x 10^{-3}) Cu - (0.611 x 10^{-6}) Cu^2 - (0.191 x live wt gain of control)

Equn. 37.1.1 All through origin. Live wt gain effect = (0.349 x 10^{-3}) Cu - (0.780 x 10^{-6}) Cu^2 - (0.151 x (live wt gain of control-mean))

Equn. 57.2 All less UK and USA Pre'65 through origin. Live wt gain effect = (0.295 x 10^{-3}) Cu - (0.696 x 10^{-6}) Cu^2 - (0.123 x (live wt gain of control-mean))

Equn. 25.7.3 Canada. Live wt gain effect = 0.0739 + (0.166 x 10^{-3}) Cu - (0.509 x 10^{-6}) Cu^2 - (0.129 x live wt gain of control)

Fig. 7. Growth response of pigs to added copper in diet:
         For various countries of origin.

These figures indicate that data from the UK and from
Canada show similar shaped curves although responses vary from
the overall equation. However, the American data from 1955 to
1975 gave a negative response to copper, although this was not
statistically significant. Further studies of these American
data indicated that since the mid 1960s, responses had tended
to be positive, whereas prior to that time, they tended to be
negative. So, the curve shown for the USA is based upon data
published in 1965 and onwards, where it can be seen that the
shape of the curve and the responses are very similar to those
of the UK and Canada. It seems almost certain that the reason
for the negative response to copper in the USA in the period
before the mid '60s was due to the absence of additional zinc
and maybe iron in the copper supplemented diets. It is an
accepted fact that increasing levels of copper can lead to zinc
deficiency symptoms in pigs if no corrective measures are taken
to increase the zinc level of the diet. Since the mid 1960s,
the Americans have been adding extra zinc and, as can be seen,
since then they have been obtaining overall positive responses
to copper.

A curve in Figure 7 illustrates the response when the UK
and the early USA data are excluded. This indicates that the
overall response to copper is not being unduly influenced by
the high numbers of UK data included. It can, therefore, be
stated that where copper has been added to pig diets in con-
junction with zinc and iron, an overall positive response has
been recorded.

OTHER FACTORS

Other factors, such as sex of pig, breed of pig and other
additives included in the diet, were studied. Unfortunately,
due to either lack of information in the references, or the fact
that the pigs were not fed as separate sexes, it was impossible
to obtain enough data to see whether sex or breed affected the
response.

As far as the additive study was concerned, scarcity of data prevented investigation of individual additives and their effects on the response to copper, but overall there were enough data from trials where other growth promoters had been included to enable the production of curves. These indicated that the addition of other growth promoters to the diet has no effect upon the response to copper. In other words, the action of copper is different from that of other growth promoters, therefore enabling one to obtain a positive response both to copper and another growth promoter.

One final factor which was considered was whether results from a few trials where more than one level of supplemented copper was used gave a different response curve. Obviously whilst these experiments are few in number, they should give more accurate measures of response. Table 2 indicates that the optimal levels of copper within these experiments with different weight ranges of pigs were almost identical and certainly not statistically significantly different from the data obtained from all the experiments, thus adding further confidence to the results shown by the overall regression equation that was ill-ustrated in the first figure.

TABLE 2

EXPERIMENTS INVOLVING MORE THAN ONE LEVEL OF COPPER SUPPLEMENTATION

	Optimal Cu level mg/kg		
Liveweight range	4 - 50 kg	4 - 60 kg	4 - 100 kg
All data	251	246	224
Experiments with multiple Cu levels	247	248	218

CONCLUSION

The data show that there is a beneficial response in both growth rate and feed conversion efficiency to the addition of copper to the diet of growing pigs. The optimal level of copper for pigs of 4 - 100 kg liveweight is 224 mg/kg, although at least 97.5% of the maximum response is obtained at levels of 188 - 260 mg/kg.

ACKNOWLEDGEMENT

I wish to acknowledge the financial assistance of the following:-

UKASTA

McKechnie Chemicals Ltd.

International Copper Research Association

and the assistance of Dr. Braude and his colleagues in extracting data from the numerous references cited.

SURVEY FOR UKASTA ON THE RESPONSE OF GROWING PIGS
TO DIETARY COPPER SUPPLEMENTATION

LIST OF REFERENCES

(This list includes all the references considered. Superscripts
following the reference number indicate: [a]reference dealt with in
full; [b]reference rejected due to lack of relevant data or duplication
of data covered by another reference; [c]full paper not found, limited
data included from abstract; [d]full paper not found, no data given
in abstract.)

1[b]   Adam, J.L., 1974. Effects of nitrovin supplementation and use of
        pure and F_1 crossbred boars on the post-weaning performance
        of growing pigs. N.Z. Jl. exp. Agric., 2, 409.

2[a]   Allcroft, R., Burns, K.N. and Lewis, G., 1961. Effect of high levels
        of copper in rations for pigs. Vet. Rec., 73, 714.

3[b]   Allee, G.L. and Hines, R.H., 1972. Supplemental copper for growing-
        finishing swine. Kansas State Univ. Swine Industry Day Rep.
        (193), 25.

4[a]   Allen, M.M., Barber, R.S., Braude, R. and Mitchell, K.G., 1958.
        Copper and zinc supplements for fattening pigs. Proc. Nutr.
        Soc., 17, xii.

5[a]   Allen, M.M., Barber, R.S., Braude, R. and Mitchell, K.G., 1961.
        Further studies on various aspects of the use of high-copper
        supplements for growing pigs. Br. J. Nutr., 15, 507.

6[a]   Amer, M.A. and Elliot, J.I., 1973. Influence of supplemental dietary
        copper and vitamin E on the oxidative stability of porcine
        depot fat. J. Anim. Sci., 37, 87.

7[a]   Anastasijevic, V., 1972. Effect of adding copper sulphate and
        Nitrosal to feeds for fattening pigs. Krmiva, 14, 241.

8[b]   Anastasijevic, V., Braude, R. and Rowell, J.G., 1963. The effect
        of the age of the pig and season of the year on the response
        to antibiotic and copper sulphate. Proc. Nutr. Soc., 22, 26.

9[b]   Baker, D.E., 1974. Copper: soil, water, plant relationships. Fedn
        Proc. Fedn Am. Socs exp. Biol., 33, 1188.

10[a] Barber, R.S., Bowland, J.P., Braude, R., Mitchell, K.G. and
Porter, J.W.G., 1961. Copper sulphate and copper sulphide
(CuS) as supplements for growing pigs. Br. J. Nutr., 15,
189.

11[a] Barber, R.S., Braude, R., Chamberlain, A.G. and Mitchell, K.G.,
1962. Effect of feeding copper sulphate to bacon pigs for
different intervals during the growing period. VIIIth int.
Congr. Anim. Prod., Hamburg, 3, 86.

12[a] Barber, R.S., Braude, R., Hosking, Z.D. and Mitchell, K.G., 1960.
Studies on liquid milk for growing pigs. The effect of its
abrupt removal from the diet and of supplementation with copper
sulphate or amino acids. Anim. Prod., 2, 105.

13[a] Barber, R.S., Braude, R. and Mitchell, K.G., 1955a. Antibiotic
and copper supplements for fattening pigs. Br. J. Nutr.,
9, 378.

14[b] Barber, R.S., Braude, R. and Mitchell, K.G., 1955b. Effect of adding
copper to the diet of suckling pigs on creep meal consumption
and liveweight gain. Chem. and Ind.,(48) 1554.

15[a] Barber, R.S., Braude, R. and Mitchell, K.G., 1960. Further studies
on antibiotic, copper and zinc supplements for growing pigs.
Br. J. Nutr., 14, 499.

16[a] Barber, R.S., Braude, R., and Mitchell, K.G., 1962a. Copper sulphate
and molasses distillers dried solubles as dietary supplements
for growing pigs. Anim. Prod., 4, 233.

17[b] Barber, R.S., Braude, R. and Mitchell, K.G., 1962b. Effect of
including an antimycotic (Griseofulvin) in the diet of growing
pigs. Anim. Prod., 4, 253.

18[a] Barber, R.S., Braude, R. and Mitchell, K.G., 1964. Further studies
on copper sulphate and molasses distillers dried solubles
as growth stimulants in the diet of growing pigs. Proc. Nutr-
. Soc., 23, 32, abs.

19[a] Barber, R.S., Braude, R. and Mitchell, K.G., 1971. Arsanilica acid,
sodium salicylate and bromide salts as potential growth
stimulants for pigs receiving diets with and without copper
sulphate. Br. J. Nutr., 25, 381.

20[a] Barber, R.S., Braude, R. and Mitchell, K.G., 1974. A note on further
studies on sodium salicylate as a growth stimulant for growing
pigs receiving diets with or without copper sulphate. Anim.
Prod., 18, 219.

21[a]   Barber, R.S., Braude, R., Mitchell, K.G. and Cassidy, J., 1955.
High copper mineral mixture for fattening pigs. Chem. and
Ind., p. 601.

22[a]   Barber, R.S., Braude, R., Mitchell, K.G., Harding, J.D.J., Lewis, G.
and Loosmore, R.M., 1968. The effects of feeding toxic ground-
nut meal to growing pigs and its interaction with high copper
diets. Br. J. Nutr., 22, 535.

23[a]   Barber, R.S., Braude, R., Mitchell, K.G. and Newport, M.J., 1967.
Response of growing pigs to copper supplementation of rations
containing dried skim milk. Anim. Prod., 9, 272, abs. 18.

24[a]   Barber, R.S., Braude, R., Mitchell, K.G., Rook, J.A.F. and
Rowell, J.G., 1957. Further studies on antibiotic and copper
supplements for fattening pigs. Br. J. Nutr., 11, 70.

25[a]   Bass, B., McCall, J.T., Wallace, H.D., Combs, G.E., Palmer, A.Z.
and Carpenter, J.E., 1956. High level copper feeding of
growing-fattening swine. J. Anim. Sci., 15, 1230, abs.

26[a]   Beames, R.M., 1969. Further trials on the response of early-weaned
pigs and rats to diets supplemented with tylosin and high
levels of copper. J. Anim. Sci., 29, 573.

27[b]   Beames, R.M. and Lloyd, L.E., 1964. Influence of soyabean meal
upon the response to copper. J. Anim. Sci., 23, 1206, abs. 38.

28[a]   Beames, R.M. and Lloyd, L.E., 1965. Response of pigs and rats to
rations supplemented with tylosin and high levels of copper.
J. Anim. Sci., 24, 1020.

29[a]   Bekaert, H., Eeckhout, W. and Buysse, F., 1967. (Effect of $CuSO_4$,
CuO, the size of the $CuSO_4$ granule and of supplement of Zn
on fattening and the Cu content of liver in fattening pigs.)
Revue Agric., Brux., 20, 1571.

30[a]   Bellis, D.B., 1961. Supplementation of bacon pig rations by
aureomycin and two levels of copper sulphate. Anim. Prod.,
3, 89.

31[d]   Bellis, D.B., 1975. A note on copper supplementation of bacon pig
diets based on fishmeal or on extracted soyabean meal. Rhod.
J. agric. Res., 13, 9.

32[a]   Berek, G., Urbanyi, L. and Lakatos, T., 1967. A note on copper
sulphate in the diet of pigs. Anim. Prod., 9, 421.

33[a]   Bowland, J.P., 1958. Antibiotic feed supplements and other additives
for swine rations. 37th Ann. Feeders' Day, Univ. Alberta,
p. 8.

34[a]   Bowler, R.J., Braude, R., Campbell, R.C., Craddock-Turnbull, J.N.,
Fieldsend, H.F., Griffith, E.K., Lucas, I.A.M., Mitchell, K.G.,
Nickalls, N.J.D. and Taylor, J.H., 1955. High copper mineral
mixture for fattening pigs. Br. J. Nutr., 9, 358.

35[b]   Braude, R., 1945. Some observations on the need for copper in diet
of fattening pigs. J. agric. Sci., Camb., 35, 163.

36[a]   Braude, R., Mitchell, K.G., Newport, M.J. and Pittman, R.J., 1970.
Response to different levels of supplementation with copper
sulphate of diets for growing pigs. Proc. Nutr. Soc., 29,
10A.

37[a]   Braude, R., Mitchell, K.G. and Pittman, R.J., 1971. An apparent
synergy between copper sulphate and Payzone in improving the
performance. Proc. Nutr. Soc., 30, 11A.

38[a]   Braude, R., Mitchell, K.G. and Pittman, R.J., 1972. Feed additive
augments response to copper. Pig Fmg., 20, (6), 34.

39[a]   Braude, R., Mitchell, K.G. and Pittman, R.J., 1973. A note on
cuprous chloride as a feed additive for growing pigs. Anim.
Prod., 17, 321.

40[a]   Braude, R. and Ryder, K., 1973. Copper levels in diets for growing
pigs. J. agric. Sci., Camb., 80, 489.

41[a]   Braude, R., Townsend, M.J., Harrington, G. and Rowell, J.G., 1962.
Effect of oxytetracycline and copper sulphate, separately
and together in the rations of growing pigs. J. agric. Sci.,
Camb., 58, 251.

42[b]   Brooks, C.C., Davis, J.W., Gidsey, R.M., Thomas, H.R. and
Meacham, T.N., 1968. Effect of antibiotics fed to pigs under
stress. Virginia Polytechnic Inst., Res. Div. Bull. (26),
17 pp.

43[a]   Brown, V.L., Warren, W.M. and Ruffin, B.G., 1969. The addition
of copper and antibiotic to rations of growing pigs. J. Anim.
Sci., 28, 141, abs. 68.

44[a]   Bugdol, G. and Schroder, H., 1968. (Copper sulphate and mixtures
of active substances for fattening pigs.) Jb. Tierernahr.
Futterung, 1967/68, 6, 294.

45[a] Bunch, R.J., McCall, J.T., Hays, V.W. and Speer, V.C., 1964. Effects
        of manganese, estrogen therapy and source of copper on response
        to high levels of copper. J. Anim. Sci., 23, 869, abs. 93.

46[a] Bunch, R.J., McCall, J.T., Speer, V.C. and Hays, V.W., 1962. Effect
        of copper supplementation on metabolism and storage of protein
        and minerals. J. Anim. Sci., 21, 989, abs. 92.

47[b] Bunch, R.J., McCall, J.T., Speer, V.C. and Hays, V.W., 1963. Effect
        of protein source, estrogen or manganese on the response to
        copper in baby pigs. J. Anim. Sci., 22, 1118, abs. 42.

48[a] Bunch, R.J., Speer, V.C. and Hays, V.W., 1960. Effects of copper
        oxide, copper sulphate and chlortetracycline on performance
        of baby pigs. J. Anim. Sci., 19, 1252, abs. 95.

49[a] Bunch, R.J., Speer, V.C., Hays, V.W., Hawbaker, J.H. and
        Catron, D.V., 1961. Effects of copper sulphate, copper oxide
        and chlortetracycline on baby pigs performance. J. Anim.
        Sci., 20, 723.

50[a] Bunch, R.J., Speer, V.C., Hays, V.W. and McCall, J.T., 1961. Effect
        of high levels of copper and chlortetracycline on performance
        of pigs. J. Anim. Sci., 20, 927, abs.

51[a] Bunch, R.J., Speer, V.C., Hays, V.W. and McCall, J.T., 1963. Effect
        of high levels of copper and chlortetracycline on performance
        of pigs. J. Anim. Sci., 22, 56.

52[a] Buysse, F. and Martin, J., 1960. (Value of large doses of copper
        as growth stimulant in fattening pigs.) Revue Agric. Brux.,
        13, 1021.

53[d] Buysse, F. and Martin, J., 1960. (The value of tetra-alkylammonium
        sterate (Dynafac) as growth stimulant in fattening pigs.)
        Revue Agric. Brux., 13, 1055.

54[a] Castell, A.G., Allen, R.D., Beames, R.M., Bell, J.M., Belzile, R.,
        Bowland, J.P., Elliot, J.I., Ihnat, M., Larmond, E.,
        Mallard, T.M., Spurr, D.T., Stothers, D.T., Wilton, S.B. and
        Young, L.G., 1975. Copper supplementation of Canadian diets
        for growing-finishing pigs. Can. J. Anim. Sci., 55, 113.

55[a] Castell, A.G. and Bowland, J.P., 1968. Supplemental copper for
        swine: growth, digestibility and carcass measurements. Can.
        J. Anim. Sci., 48, 403.

56[d] Chiomba, R.A., 1969-70. Some effects of high-copper-supplementation
       of diets for baby pigs over the period 2 - 28 days of age.
       M.Sc. Thesis, Univ. Bristol.

57[a] Clawson, A.J. and Alsmeyer, W.L., 1973. Chemotherapeutics for pigs.
       J. Anim. Sci., 37, 918.

58[a] Combs, G.E., Ammerman, C.B., Shirley, R.L. and Wallace, H.D., 1966.
       Effect of source and level of dietary protein on pigs fed
       high copper rations. J. Anim. Sci., 25, 613.

59[a] Corzo, M., Hays, V.W., Cromwell, G.L. and Kratzer, D.D., 1972.
       Related effects of dietary copper, sulphide and molybdenum
       for pigs. J. Anim. Sci., 35, 215, abs. 191.

60[a] Cosic, H., 1972. (Effectiveness of several copper compounds in
       the ration of early weaned pigs.) Agron. Glasn., 34, 215.

61[d] Crnojevic, Z., Seles, J., Jancic, S., Zlatic, H. and Pesut, M.,
       1970. (Supplementation of rations with copper sulphate and
       Galofac on performance of growing-finishing pigs.) Symp.
       Pig Prod. Anim. Nutr., Zagreb, p. 133.

62[a] Dammers, J. and Van der Grift, J., 1959. (Comparative feeding trials
       with different quantities of copper sulphate.) Versl.
       Landbouwk. Onderz., Wageningen, No. 65.12, 7.

63[a] Degoey, L.W., Wahlstrom, R.C. and Emerick, R.J., 1971. Studies
       of high level copper supplementation to rations for growing
       swine. J. Anim. Sci., 33, 52.

64[a] Drouliscos, N.J., Bowland, J.P. and Elliot, J.I., 1970. Influence
       of supplemental dietary copper on copper concentration of
       pig blood, selected tissues and digestive tract contents.
       Can. J. Anim. Sci., 50, 113.

65[c] Dzilinski, E., Reka, J. and Raslawski, Z., 1969. (Use of copper
       sulphate for growing pigs.) Przegl. hodowl., 37, (8), 10.

66[a] Fagan, V.J., Iles, R.D., Slowitzky, Z. and Brocksopp, R.E., 1961.
       Some observations on the high level copper supplementation
       of pig rations. J. Agric, Sci., Camb., 56, 161.

67[c] Filar, J., 1965. (Effect of $CuSO_4$ on weight gains and feed
       utilisation in fattening pigs.) Inst. Zootech., Krakow,
       Wydawn. wlasne, No. 185, 67.

68[b] Galik, R., 1969. (Effect of different levels of cupric sulphate
       in rations of fattening pigs on digestibility of organic
       nutrients.) Zivocisna vyroba, 14, 729.

69[c] Gavrila, V., Jurubescu, V., Moldoveanu, X., Popescu, V. and
       Mihailescu, P., 1967. (Influence of copper sulphate on the
       productivity of swine.) Lucr. stiint. Inst. agron. N.
       Balcescu, Ser. C, X, 157.

70[b] Giessler, H. and Kirchgessner, M., 1959. (Effects of copper and
       manganese additives in pig fattening and chick rearing.)
       Landw. Forsch., 12, 159.

71[a] Gipp, W.F., Pond, W.G. and Smith, S.E., 1967. Effects of dietary
       copper, molybdenum, sulphate and zinc on body weight gain,
       hemoglobin and liver copper storage of growing pigs. J. Anim.
       Sci., 26, 727.

72[a] Gipp, W.F., Pond, W.G. and Walker, E.F., 1973. Influence of diet
       composition and mode of copper administration on the response
       of growing-finishing swine to supplemental copper. J. Anim.
       Sci., 36, 91.

73[a] Grashuis, J., 1957. (Dangers of use of antibiotics in mixed feeds.)
       Tijdschr. Diergeneesk., 82, 775.

74[a] Grzwinski, L., Grzegorzak, A. and Pres, J., 1972. (Effect of
       de-worming, addition of copper sulphate and darkness in
       piggeries on performance of growing pigs.) Medycyna wet.,
       28, 345.

75[a] Gupta, S., Moulick, S.K. and Bhattacharya, S., 1964. Effect of
       high-level copper supplementation in the ration of growing
       pigs. Emp. J. exp. Agric., 32, 331.

76[a] Hansen, V. and Bresson, S., 1975. Copper sulphate as a feed additive
       to bacon pigs. Acta Agric. scand., 25, 30.

77[a] Hawbaker, J.A., Speer, V.C., Hays, V.W. and Catron, D.V., 1961.
       Effect of copper sulphate and other chemotheraputics in growing
       swine rations. J. Anim. Sci., 20, 163.

78[b] Hawbaker, J.A., Speer, V.C., Jones, J.D., Hays, V.W. and
       Catron, D.V., 1969. Effect of copper sulphate and antibiotics
       on growth rate, feed conversion and fecal flora of growing
       pigs. J. Anim. Sci., 18, 1505, abs. 116.

79[a] Hays, V.W., Cromwell, G.L. and Overfield, R.O., 1971. Effect of
       copper and vitamin E on response of pigs fed corn and wheat
       base diets. J. Anim. Sci., 33, 1149, abs. 52.

80[b] Hedges, J.D. and Kornegay, E.T., 1973. Interrelationship of dietary copper and iron as measured by blood parameters, tissue stores and feed lot performance of swine. J. Anim. Sci., 37, 1147.

81[a] Hennig, A., Gruhn, I. and Anke, M., 1968. (Copper sulphate for fattening pigs. 4. Effect of Cu supplements on crude nutrients, Cu, Fe, Zn and Mo in some body compounds and the faecal flora.) Jb. Tierernahr. Futterung, 1967/68, 6, 278.

82[a] Ho, S.K., Elliot, J.I. and Jones, G.M., 1975. Effects of copper on performance, fatty acid composition of depot fat and fatty acyl desaturase activities in pigs fed a diet with or without supplemental copper. Can. J. Anim. Sci., 55, 587.

83[c] Isakov, D., 1962. (Effect of copper sulphate and antibiotics on weight gains and feed utilisation in fattening pigs.) Vet. Glasn., 16, 1191.

84[b] Iwanska, S., 1973. (Copper metabolism in piglets.) Zesz. nauk. Akad. roln techn. Olsztynie (110), Zootechnica, 3, 29 pp.

85[b] Jancic, S., Crnojevic, Z., Pesut, M., Crnojevic, T. and Pozezanac, T., 1973. (The influence of source of high copper levels in diets on Cu, Fe, Mn and vitamin A concentration in some tissues of fattening pigs.) Poljopr. znanst. Smotra, 30, 353.

86[a] Jucker, H., 1961. (Effect of copper sulphate supplements to the feed of growing pigs on live weight gain, feed utilisation and copper content of several tissues.) Mitt. Lebensmittelunters. Hyg., Bern, 52, 580.

87[a] Kellogg, T.F., Quinn, L.Y., Hays, V.W., Catron, D.V. and Speer, V.C., 1960. Effect of chemotherapeutics on the fecal flora of young pigs. J. Anim. Sci., 19, 1270, abs. 140.

88[a] King, J.O.L., 1960. The effect of environmental temperature on the response of growing pigs to dietary supplements of an antibiotic and copper sulphate. Vet. Rec., 72, 304.

89[a] King, J.O.L., 1963. The effect of water intake on the efficiency of copper sulphate as a growth stimulant for pigs. Vet. Rec., 75, 651.

90[a] King, J.O.L., 1964. The effect of the level of protein in the diet on the efficacy of copper sulphate as a growth stimulant for pigs. Br. Vet. J., 120, 531.

91[a] King, J.O.L., 1969. Influence of the water treatment of rations
       supplemented with copper sulphate and an antibiotic on the
       growth rate of pigs. Expl Husb., (18), 25.

92[b] Kirchgessner, M. and Grassmann, E., 1970. The dynamics of copper
       absorption. In: 'Trace element metabolism in animals'.
       p. 277. edited by Mills, C.F. London & Edinburgh: Livingstone.

93[a] Kleeman, J., Powelleit, G., Hennig, A., Heuschkel, M. and Viewig, S.
       1966. (The use of copper sulphate in pig fattening. 3.)
       Jb. Tierernahr. Futterung, 5, (1964/65), 231.

94[a] Kline, R.D., Corzo, M.A., Hays, V.W. and Cromwell, G.L., 1973.
       Related effects of copper, molybdenum and sulfide on per-
       formance, hematology and copper stores of growing pigs. J.
       Anim. Sci., 37, 936.

95[a] Kline, R.D., Hays, V.W. and Cromwell, G.L., 1971. Effects of copper
       molybdenum and sulfate on performance, hematology and copper
       stores of pigs and lambs. J. Anim. Sci., 33, 771.

96[a] Kline, R.D., Hays, V.W. and Cromwell, G.L., 1972. Related effects
       of copper, zinc and iron on performance, hematology and copper
       stores of pigs. J. Anim. Sci., 34, 393.

97[a] Kornegay, E.T. and Thomas, H.R., 1973. Feed additives for swine.
       2. Copper and iron additions for grower rations. Virginia
       Polytechnic Inst. Res. Div. Rep. 149, 5 pp.

98[a] Kornegay, E.T., Thomas, H.R. and Kramer, C.Y., 1975. Effect on
       subsequent feedlot performance of rotating or withdrawing
       dietary antibiotics from swine growing and finishing rations.
       J. Anim. Sci., 41, 1555.

99[a] Kosanovic, M., Zivkovic, S. and Nikolic, M., 1967. (Antibiotics,
       Cu and As as stimulants in nutrition of fattening pigs.)
       Veterinaria, Saraj., 16, 339.

100[d] Kovalenko, N.A., 1965. (Effect of Cu on the productivity of pigs.)
       Svinovodstvo, 19, (12), 33.

101[c] Lalov, N., Enchev, S., Gurkov, G. and Mitreva, V., 1967. (The use
       of copper sulphate as a stimulant in pig fattening.) Vet.
       Med. Nauki, Sof., 4, (6), 91.

102[a] Lavorenti, A., 1972. Utilisation of dietary iron by pigs and rats.
       Diss. Abstr. Int., 33, 509-b.

103[c] Leroch, Z., 1963. (Influence of copper sulphate on weight gain
        and feed efficiency in fattening pigs.) Zesz. nauk. W.S.R.
        Wroclaw, Zootech., 11, (52), 189.

104[b] Livingstone, R.M., 1967. The supplementation of diets for early
        weaned pigs with copper and antibiotics. Proc. Holmenkollen
        Symp. on 'Antibiotics in animal nutrition'.

105[a] Livingstone, R.M. and Livingstone, D.M.S., 1968. Copper sulphate
        and antibiotics as feed additives for early weaned pigs.
        J. Agric. Sci., Camb., 71, 419.

106[a] Lucas, I.A.M. and Calder, A.F.C., 1957a. A comparison of five levels
        of copper sulphate in rations for growing pigs. Proc. Nutr.
        Soc., 16, i.

107[a] Lucas, I.A.M. and Calder, A.F.C., 1957b. Antibiotics and a high
        level of copper sulphate in rations for growing pigs. J.
        Agric. Sci., Camb., 49, 184.

108[a] Lucas, I.A.M., Livingstone, R.M. and Boyne, A.W., 1962. Copper
        sulphate as a growth stimulant for pigs: effect of composition
        of diet and level of protein. Anim. Prod., 4, 177.

109[a] Lucas, I.A.M., Livingstone, R.M., Boyne, A.W. and McDonald, I.,
        1962. The early weaning of pigs. VIII. Copper sulphate
        as a growth stimulant. J. Agric. Sci., Camb., 58, 201.

110[a] Lucas, I.A.M., Livingstone, R.M. and McDonald, I., 1961. Copper
        sulphate as a growth stimulant for pigs: effect of level and
        purity. Anim. Prod., 3, 111.

111[a] Matre, T., 1971. (Experiments on supplementation of copper to
        rations for bacon pigs.) Norges Landbrukshogskole, Inst.
        husdyrernaer. foringslaere Beretn. nr. 142, 29 pp.

112[a] Meyer, H. and Kroger, H., 1973. (Comparative studies on piglets
        on growth promoting effects of copper and antibiotics.)
        Zuchtungskunde, 45, 439.

113[d] Miller, H.W., 1969. The value of new feed additives for
        growing-finishing swine. Diss. Abstrs, 30, 2482-B.

114[a] Milosavljevic, S. and Sovljanski, B., 1959. (Influence of copper
        in the feed on weight gains and feed utilisation in pigs.)
        Vet. Glasn., 13, 505.

115[a] Milosavljevic, S., Sovljanski, B. and Pavlovic, S., 1962.
(Comparative examinations of the effect of copper sulphate
and antibiotics supple ments on weight gain and feed conversion
of fattening pigs.) Vet. Glasn., 16, 493.

116[d] Milosavljevic, S., Sovljanski, B. and Pejin, R., 1960. (Influence
of copper sulphate in the feed on growth and feed utilisation
in pigs. 1.) Vet. Glasn., 14, 487.

117[a] Milosavljevic, S., Sovljanski, B. and Pejin, R., 1960. (Influence
of copper sulphate in the feed on growth and feed utilisation
in pigs. 2.) Vet. Glasn., 14, 841.

118[a] Moulick, S.K., Bhattacharya, S. and Gupta, S., 1965. Effects of
terramycin, aureomycin and high level of copper sulphate on
growing pigs. Indian J. Vet. Sci., 35, 275.

119[a] Myres, A.W. and Bowland, J.P., 1972. Effects of high levels of
dietary copper on endogenous lipid metabolism in the pig.
Can. J. Anim. Sci., 52, 113.

120[a] Myres, A.W. and Bowland, J.P., 1973. Effects of environmental
temperature and dietary copper on growth and lipid metabolism
in pigs. I. Growth, carcass quality, and tissue copper
levels. Can. J. Anim. Sci., 53, 115.

121[a] Myres, A.W. and Bowland, J.P., 1975. Influence of dietary copper
on the fatty acid composition of adipose tissue lipids and
on the level and composition of plasma free fatty acids in
growing pigs fed individually or in groups. Can. J. Anim.
Sci., 55, 315.

122[a] NCR-42 Committee, 1974. Co-operative regional studies with growing
swine: effects of vitamin E and levels of supplementary copper
during the growing-finishing period on gain, feed conversion
and tissue copper storage in swine. J. Anim. Sci., 39, 512.

123[d] Obenko, K.S. and Ganza, A.A., 1970. (Trace elements for young
pigs.) Svinovodstvo, 24, (11), 15.

124[a] Omole, T.A., Adebayo, A.A. and Ilori, J.O., 1974. The effects of
supplemental dietary copper on the performance of growing
pigs in the tropics. Nutr. Rep. Int., 10, 235.

125[a] Omole, T.A. and Bowland, J.P., 1974. Copper and zinc supplementation
of pig diets containing soyabean meal or rapeseed meal
(Brassica campestris cv. Span). Can. J. Anim. Sci., 54, 363.

126[a] Omole, T.A. and Bowland, J.P., 1974. Copper, iron and manganese
        supplementation of pig diets containing either soyabean meal
        or low gluco sinolate rapeseed meal. Can. J. Anim. Sci.,
        54, 481.

127[a] Rerat, A., 1971. (Growth characteristics and body composition of
        pigs as influenced by addition of copper to the diet.)
        Journees de la Recherche Porcine en France, I.N.R.A., I.T.P.,
        p. 167.

128[a] Parker, G.R., Prince, T.J., Cromwell, G.L., Hays, V.W. and
        Kratzer, D.D., 1975. Effects of Ca-P and Cu levels on
        performance and bone traits of G-F swine. J. Anim. Sci.,
        41, 325, abs.

129[a] Parris, E.C.C. and McDonald, B.E., 1969. Effect of dietary protein
        source on copper toxicity in early weaned pigs. Can. J. Anim.
        Sci., 49, 215.

130[a] Plumlee, M.P., Krider, J.L., Conrad, J.H., Morse, E.V., Underwood, L,
        and Lavorenti, A., 1972. Copper and iron in growing-finishing
        swine diets. J. Anim. Sci., 35, 222, abs. 220.

131[b] Podabaj, G.F. and Zorova, M.V., 1967. (Growth stimulants for
        piglets.) Svinovodstvo, 21, (1), 17.

132[a] Prince, T.J., Hays, V.W. and Cromwell, G.L., 1974. Effects of
        calcium, phosphorus and copper on performance and liver copper
        levels in pigs. J. Anim. Sci., 39, 981, abs. 66.

133[a] Ritchie, H.D., Luecke, R.W., Baltzer, B.V., Miller, E.R.,
        Ullrey, D.E. and Hoefer, J.A., 1962. Supplementation of
        normal-calcium rations for swine with chlortetracycline, zinc,
        copper oxide or copper sulphate. J. Anim. Sci., 21, 1010,
        abs. 185.

134[a] Ritchie, H.D., Luecke, R.W., Baltzer, B.V., Miller, E.R.,
        Ullrey, D.E. and Hoefer, J.A., 1963. Copper and zinc inter-
        relationships in the pig. J. Nutr., 79, 117.

135[a] Rodrigues, A.J., Velloso, L., Becker, M., Spers, A., Silveira, J.J.
        and Yamamoto, M., 1967. (The effects of copper sulphate on
        high and low energy rations for growing pigs.) Bolm. Ind.
        Anim., 23, 115.

136[a] Rubach, G. and Gruhn, K., 1963. (The effect of aminobutyric acid,
        oxy tetracycline and copper sulphate in fattening pigs on
        rations with vegetable protein feeds.) Jb. Arbeitsgemeinsch.
        Futterungsberat., 4, (1961-62), 210.

137[a] Ruszczyc, Z., Fritz, Z. and Pres, J., 1971. (Weight gains, intake
        of feed carcass quality and some blood values for fattening
        pigs given different copper salts.) Roczn Nauk roln., Ser.
        B, 93, 27.

138[a] Ruszczyc, Z. and Glaps, J., 1960. (Influence of antibiotics, copper
        sulphate and 3-nitro-hydroxyphenylarsonic acid on the weight
        gains and carcass characteristics of pigs.) Roczn. Nauk roln.,
        Ser. B, 75, 541.

139[a] Ruszczyc, Z. and Glaps, J., 1962. (Copper sulphate and oxytetra-
        cycline for fattening pigs.) Roczn. Nauk roln., Ser. B, 78,
        569.

140[a] Schurch, A., 1956. (The effect of high copper supplements on growth
        rate of fattening pigs.) Mitt. Geb. Lebensmittelunters. Hyg.,
        Bern, 47, 458.

141[a] Scott, K.W., Noland, P.R. and Heck, M.C., 1958. Effect of copper
        levels, antibiotics, and/or yeast in the ration of
        growing-finishing swine. Proc. Ass. Sth agric. Wkrs, 55th
        Ann. Convention, p. 78.

142[a] Suttle, N.F. and Mills, C.F., 1966a. Studies on the toticity of
        copper to pigs. 1. Effects of oral supplements of zinc and
        iron salts on the development of copper toxicosis. Br. J.
        Nutr., 20, 135.

143[a] Suttle, N.F. and Mills, C.F., 1966b. Studies on the toxicity of
        copper to pigs. 2. Effect of protein source and other dietary
        components on the response to high and moderate intakes of
        copper. Br. J. Nutr., 20, 149.

144[a] Teague, H.S., Grifo, A.P. and Roller, W.L., 1972. High levels of
        copper in paste feed for growing-finishing pigs. Ohio Agric.
        Res. & Develop. Center Res. Summary 61, p. 5.

145[a] Thomas, H.R. and Kornegay, E.T., 1973. Feed additives for swine.
        3. Comparison of copper and antibiotics for grower rations.
        Virginia Polytechnic Inst. Res. Div. Rep. 150, 5 pp.

146[a] Todd, A.C.E., 1965. Using more copper in feed for growing pigs.
        Qd agric. J., 91, 715.

147[c] Tschiderer, K., 1961. (Pig fattening trials with copper supple-
        ments. (Preliminary trials).) Die Bodenkultur, 12A, 74.

148[c] Vainshtein, Y.I. and Urakov, V.I., 1971. (Increased amounts of
        copper sulphate in feeds for pigs.) Khimiya sel'. Khoz., 9, 373.

149[b] Van der Wal, P., 1966. (The effect of 10 ppm penicillin-streptomycin
         on growing pigs of 20 - 90 kg liveweight.) Meded. Landb.
         Hogesch., Wageningen, 66-12, 40 pp.

150[c] Van der Wal, P. and van Weerden, E.J., 1961. (Effect of small
         amounts of penicillin in the feed on growth and feed intake
         of table poultry and fattening pigs.) Meded. Landb. Hogesch.,
         Wageningen, 61, (14), 31 pp.

151[a] Velickovic, G., Jelic, T., Stankovic, M. and Nikolic, N., 1971.
         (Effect of copper; and manganese on productive capacity of
         fattening pigs.) Stocarstvo, 25, 281.

152[a] Walker, N., Hines, W.J.W. and Elliott, R.J., 1971. The effects
         of dietary copper levels on the performance and muscle and
         fat characteristics of growing pigs. Rec. Agric. Res., N.
         Ire., 19, 53.

153[a] Wallace, H.D., McCall, J.T., Bass, B. and Combs, G.E., 1960. High
         level copper for growing-finishing swine. J. Anim. Sci.,
         19. 1153.

154[d] Wingert, F.C., 1958. Maryland Nutr. Conf., Feed Mfrs, p. 34.

155[b] Young, L.G., 1967. Copper and simplified starter rations for
         swine. J. Anim. Sci., 26, 912, abs. 118.

156[a] Young, L.G., Brown, R.G., Ashton, G.C. and Smith, G.C., 1970a.
         Effect of copper on the utilisation of raw soybeans by market
         pigs. Can. J. Anim. Sci., 50, 717.

157[a] Young, L.G. and Jamieson, J.D., 1970b. Protein and copper
         supplementation of corn-soybean meal diets for young pigs.
         Can. J. Anim. Sci., 50, 727.

158[a] Zivkovic, S., Adzic, A., Gradinac, D., Pijin, S. and Jovanovic, B.,
         1968. (Effect of Galofak SP 125, chloramphenicol, copper,
         arsanilic acid and nitrofurazone in nutrition of fattening
         pigs.) Stocarstvo, 22, 61.

159[a] Zivkovic, S., Glavaski, S., Pijin, S. and Kosanovic, M., 1967.
         (Effect of application of Galofac SP-25, furazolidone and
         copper in the nutrition of early weaned young pigs.) Savrema
         poljopr., (12), 981.

COPPER IN FEEDINGSTUFFS
POLITICAL AND LEGISLATIVE SITUATION IN THE COMMUNITY

A.D. Bird

Agricultural Trade Consultant - 'Woodbury'
10, Bromley Lane, Chislehurst, Kent, BR7 6LE, UK.

ABSTRACT

Germany, France, Ireland, Luxembourg, Netherlands and United Kingdom permit 200 mg/kg maximum copper in pig foods. Belgium, Denmark and Italy permit only 125 mg/kg due to anxiety about build up of copper in the soil. For similar reasons Germany restricts the higher level to foods for younger pigs, and the Netherlands subsidises transport of slurry away from the main centres of pig production.

The maximum amount of copper reaching the soil via slurry if all pig foods in the Community received high level copper, is estimated at 9 500 tonnes: 3 500 tonnes of this could be attributed to inclusions above the generally acceptable level of 125 mg/kg.

Earlier published work in the UK about the economics of copper inclusions at 200 mg/kg, is updated. This shows a return from copper supplementation of £3.23 per bacon pig.

It is suggested that the option to include copper in pig foods up to 200 mg/kg, be extended for five years to allow further research about the build up of copper in the soil. Also consideration should be given to controls on application of pig slurry similar to those in operation for industrial/domestic sewage sludge.

INTRODUCTION

The EEC Additives Directive 70/524 fixes the maximum content of copper in animal feedingstuffs (naturally present + added). Details are given in the following table

TABLE 1

MAXIMUM PERMITTED COPPER LEVEL IN ANIMAL FEEDING STUFFS

EEC Additives directive 70/524 reference	Species of animal	Maximum copper content mg/kg complete feedingstuffs
Annex I (mandatory in national legislation)	Pigs	125
	Other species	50
Annex II (optional in national legislation until 31 December 1980)	Pigs	200

Entries in Annex II of the Directive are temporary and subject to review, by the EEC Standing Committee for Feedingstuffs, within a specified period. The Committee has to decide whether the additive in question should be promoted to Annex I for universal usage, or deleted i.e. prohibited except on veterinary prescription. Current practice does not allow additives to remain permanently in Annex II. But, in exceptional circumstances, the period of authorisation may be extended.

The existing Annex II entry for copper, which expires on 31 December 1980, has been extended for a year on four separate occasions previously. It is expected to be extended shortly for a further year pending consideration of a report by the EEC Standing Committee on Animal Nutrition (SCAN).

This paper reviews the law regarding maximum copper inclusions in animal feedingstuffs, and political attitudes, in the individual member states; estimates the maximum quantity of copper reaching the soil through pig slurry; estimates the

economic benefits of copper supplementation of pig foods; and
suggests possible solutions which would allow pig farmers to
continue to enjoy such benefits and allay the fears of the
environmental and consumer protection lobbies.

THE LAW IN MEMBER STATES

In an answer in the European Parliament on 18 March 1978
(Official Journal of the European Communities, 24 April 1978)
the Commission gave details of the maximum copper content in
feedingstuffs fixed by the individual member states in accordance
with Directive 70/524. The author has updated this information
following visits and discussions during 1978/79 in the
individual member states. The current national situation is
believed to be as set out in Table 2.

TABLE 2

MAXIMUM COPPER CONTENT IN FEEDINGSTUFFS FIXED BY MEMBER STATES

Member State	Species of animal	Maximum copper content mg/kg complete feedingstuff
Germany	Piglets up to 16 weeks	200
	Pigs over 16 weeks	125
	Other species	50
Belgium) Denmark) Italy)	Pigs Other species	125 50
France) Ireland) Luxembourg) Netherlands) United Kingdom)	Pigs Other species	200 50

The three countries - Belgium, Denmark and Italy, which
have not taken advantage of the 200 mg/kg maximum copper content
have reached this decision on environmental grounds. Belgium
and Denmark are small countries with large concentrations of
pigs. In Italy there are heavy concentrations of pigs in
certain areas in the north viz. Lombardy, Veneto, Emilia-
Romagna, etc. The respective governments of these countries
maintain that slurry from pigs fed at the 200 mg/kg copper
level could lead to an unacceptable build up of copper in the
soil.

The Federal Republic of Germany has some doubts on
environmental grounds and has limited 200 mg/kg copper
inclusions to piglets up to 16 weeks old.

Although the Netherlands continues to permit 200 mg/kg
copper inclusions in pig foods the Government is under
increasing public pressure from the environmental lobby about
build up of copper in the soil from pig slurry, particularly
in those areas of high pig concentrations. In an effort to
control the situation the Government has subsidised the cost
of transporting slurry away from the areas where there are
high concentrations of pigs.

France is the only other country which has felt it
necessary to take special measures in respect of slurry disposal,
only then in Brittany, where there is a major concentration of
pigs. Here application of slurry is controlled, although
probably more on account of odour than to prevent build up of
copper in the soil.

There is sufficient voting strength in the three countries
that have not opted for the 200 mg/kg copper level in pig foods
to have this removed from Annex II, should the question
eventually be taken to a vote in the Standing Committee for
Feedingstuffs. There seems little prospect of these countries
changing their attitudes. The Netherlands Government, faced
with a vote, might also come out against continuing the

200 mg/kg level in view of the weight of public opinion in their country.

COPPER IN PIG SLURRY

Copper in pig foods is clearly a very controversial and emotional issue. It is essential, therefore, to establish certain basic facts; not least, an estimate of how much copper is reaching the land via the pig food route.

Unfortunately this information is not readily obtainable since the copper salts, mostly sulphate, used in pig foods, have a number of other agricultural uses, e.g. fungicides, fertilisers.

Taking the official figures for the production of compound pig foods by commercial manufacturers in the EEC (Table 3) and making certain assumptions, it is possible to estimate the maximum quantity of copper which would be consumed by the EEC pig population annually if there were universal acceptance of the 200 mg/kg level.

TABLE 3

PRODUCTION OF COMPOUND FEED FOR PIGS IN THE EEC 1978

Member State	Production million tonnes
Germany	5.52
Belgium	2.77
Denmark	1.67
Italy	1.98
France	4.70
Ireland	0.48
Luxembourg	N / A
Netherlands	5.43
United Kingdom	2.30
Total EEC	24.85

Taking the United Kingdom first, and assuming that all
compound feeds receive copper supplementation at 200 mg/kg i.e.
200 g/tonne, (breeding and creep feeds not supplemented will
be offset by higher levels in concentrate type compounds) the
calculation of total copper consumption via compound feeds is
as follows:

Total UK copper consumption
via pig compounds = 2.30 million x 0.0002 = 460 tonnes

Not all UK pig farmers use bought in manufactured compounds and some allowance must be made for copper added to
farm mixed rations for pigs. Assuming that half the UK pig
population is fed on farm mixed rations then the calculation
of total copper consumption in all pig foods is as follows:

Total UK copper consumption
via all pig rations = 460 x 2 = 920 tonnes.

This figure has been accepted by the major copper sulphate
manufacturers in the UK as a realistic one.

Applying the same assumptions to the Community as a whole
the calculation is as follows:

Total EEC copper consumption
via all pig rations = 24.85m x 0.0002 x 2 = 9 940 tonnes

It is important to remember that this figure is an
approximate estimate of the maximum amount of copper which
would be fed to pigs if all member states permitted the Annex
II maximum of 200 mg/kg in pig foods.

Assuming that the Annex I maximum of 125 mg/kg copper
content is not at risk, then the issue can be quantified still
further, on the basis of previous assumptions, to the effects
of including an extra 75 mg/kg of copper in pig foods. The
calculation of the total additional copper involved in

raising the inclusion rate from 125 to 200 mg/kg is as follows:

Total maximum EEC additional copper consumption via all
pig rations in raising inclusion rate from:
125 to 200 mg/kg = 9 940 tonnes - 6 212 (24.85 m x
0.000125 x 2) = 3 728 tonnes

How much of this copper is excreted and eventually reaches
the soil via pig slurry? Wilson et al. (1979) quote the
environmentalists' view that 95% of all added copper in a pig's
diet is voided. On this basis the maximum total quantity of
copper reaching the soil via pigs can be rounded down to about
9 500 tonnes, and the maximum total additional copper within
this total, attributable to higher level inclusions, to about
3 500 tonnes. In practice these totals will be considerably
less, but at least they should help to put the problem in some
sort of perspective.

INDUSTRIAL INVOLVEMENT

In terms of copper sulphate, the principal form in which
copper is fed to pigs, we are talking about four times the
weight of copper. Even so, 39 760 tonnes (9 940 x 4) of copper
sulphate is only a small proportion of total EEC usage for
agricultural and industrial purposes. The chemical industry
cannot be accused, therefore, of reckless exploitation of
copper usage in pig foods for large scale profits regardless
of environmental considerations. Neither is the compound
feed industry's attitude a solely selfish one, other than in
the sense that by contributing towards greater profitability
it is enabling its customers to stay in business.

OTHER SOURCES OF COPPER

Pig slurry is not the only route through which copper
reaches the soil. The spreading of industrial/domestic sewage
sludge and the direct application to soil, herbage and crops
are others. It is unfortunate that pig foods should be singled

out as the major threat to the environment.

Wilson et al. (1979) say that copper levels in mixed industrial/domestic sewage sludge are just as high as those in pig slurry. They go on to suggest that to alleviate any public concern a Code of Practice detailing the correct way to spread pig slurry on agricultural land is necessary. This has already been done in the UK in respect of sewage sludge (Report of the Working Party on the Disposal of Sewage Sludge, National Water Council, Department of the Environment, 1, Queen Anne's Gate, London, 1977).

ECONOMICS OF COPPER INCLUSIONS

The main beneficiary from high level copper inclusions in pig foods is not the copper sulphate manufacturer, not the feed compounder but the farmer. Wilson et al. (1979) made an evaluation of the economics of copper additions to pig foods. This showed that at 200 mg/kg copper inclusion was highly beneficial and gave a growth rate increase of 6.5%. Assuming there are 134 days of copper supplementation needed before a pig reaches bacon weight (90 kg) at an inclusion rate of 200 mg/kg the extra value of each pig is £3.30. The cost of the copper to produce this added benefit is only 7p, thus giving a return from copper supplementation of £3.23 per pig. (These figures have been revised on the basis of present day costings).

The inclusion of copper in pig foods at 200 mg/kg has become an accepted practice in the United Kingdom over the past 25 years. If the maximum were reduced to 125 mg/kg then it is likely that veterinary practitioners would be asked to prescribe at the old maximum level. The net result would be the same quantity of copper being consumed but at higher cost and reduced profit to pig farmers, since under UK medicated feed law, feeds containing prescription only additives cannot be manufactured and stored in advance of demand, thus increasing unit cost of production.

CONCLUSIONS

The political dilemma facing the Commission is how to
reconcile the economic benefits to farmers of copper supplementation of pig foods at 200 mg/kg with possible long term
environmental hazard from build up of copper in the soil from
pig slurry. The continual extension of the validity of the
entry for copper in Annex II of Directive 70/524 over the past
five years is an encouraging sign. Clearly the Commission is
not going to be rushed into a hasty decision until it has had
an opportunity to examine all the facts.

The weight of public opinion against copper in some
member states, however ill-founded, is such that there is no
prospect of getting the 200 mg/kg level in pig foods adopted
universally throughout the Community i.e. promoted to Annex I
of Directive 70/524. But surely this is a situation which
calls not for a uniform solution, but for continued discretion
by the member states in the light of national circumstances
e.g. size/concentration of pig population; any copper deficiency
in the soil; etc. In the United Kingdom, for example, where
there is widespread deficiency of copper in the soil, it would
be ludicrous to deny pig producers the right to continue to be
able to buy pig foods containing copper at 200 mg/kg for no
better reason than pig slurry disposal problems in Flanders or
Lombardy. This would be harmonisation for harmonisation's sake
and a negation of the principles of the Rome Treaty.

The existing entry for copper in Annex II of Directive
70/524 provides the ideal mechanism for national discretion
except that the authority is terminal after a relatively short
period. If the option to continue to use 200 mg/kg in pig
foods were extended for a longer period say up to five years,
this would at least ensure that further research could be
conducted to establish the extent to which there has been a
build up of copper in the soil in those countries where copper
supplementation has been a feature of animal nutrition in the
past 25 years. At the same time it would remove, at least for

the time being, the constant threat to economic pig production in a large section of the Community.

REFERENCE

Wilson, P.N., Brigstocke, T.D.A. and Cooke, B.C. 1979. Copper as an inexpensive growth promoter for the pig. 'Process Biochemistry' August 1979 (Wheatland Journals Ltd., 157 Hagden Lane, Watford, Herts.)

DISCUSSION

R.D. Davis *(UK)*

I have a small observation to make on the comparison of additions of copper to agricultural land from animal wastes and sludge. Mr. Bird calculates, for the UK, 920 t of copper from animal wastes; the conservative amount for sewage sludge is about 300 t.

R. Braude *(UK)*

Although, Mr. Bird assumes that all pigs are receiving copper, a more likely assumption is that half or less receive it.

A.D. Bird *(UK)*

In my presentation, I explained that the maximum was used to make sure that we saw the possible extent of the prob- lem. However, I agree that the figure is higher than in actual practice.

A. Dam Kofoed *(Denmark)*

I would like to ask Dr. Bories, what is the real physio- logical effect of 200 - 250 mg/kg of copper in the feed on the pig? We‘ have had a lot of discussion but we have not had a definitive answer to this question.

G. Bories *(France)*

I am afraid I cannot answer, I am not a physiologist.

A. Madsen *(Denmark)*

The answer depends on the form of copper, with copper sulphate, the liver contained 256 ppm Cu but with the same amount in the form of copper oxide, only one tenth of this amount was present. This raises the question, are we concerned with copper, or the form in which it is administered such as copper sulphate or copper oxide. If we changed to copper oxide instead of copper sulphide, is there then no risk?

G. Bories

Is copper oxide efficient as a growth promoter for pigs? Copper sulphate is probably one of the best forms. In Germany or Switzerland there is a proposal to use an organic form of copper such as copper methionine which is claimed to have the same efficiency as copper sulphate but with an administered copper level of about 80 ppm.

R. Braude

The difficulty with copper oxide is that the fate of copper given in this form is not known.

FINAL SESSION

Chairman: J.K.R. Gasser

TELEGRAM

J.K.R. Gasser *(UK)*

At this point, I will introduce the telegram that has
been received from Madame Dr. S. Dormal-van den Bruel, who is
the Secretary of the Animal Nutrition Committee. In the telegram Dr. Dormal informed the Workshop that the EEC Scientific
Committee for Animal Nutrition was set up by Commission
decision 79/791/EEC of 24th September, 1976 (OJ No. L 279 of
9th October, 1976) to provide the Commission with informed opinion on matters relating to animal nutrition and stock farming
and the effects of production techniques on food quality and
the environment.

Dr. Dormal continued,

"The opinions expressed by the Committee have permitted
the Commission to base its proposals of Directives in the field
of additives and undesirable substances in feedingstuffs on
sound scientific background evidence. With regard to copper,
in accordance with the provisions of Council Directives concerning additives in feedingstuffs, as last amended by the 23rd
Commission Directive of 4th July, 1980, the addition of copper
salts is authorised at Community level up to 125 mg of copper
in complete feedingstuffs for pigs, and 50 mg/kg for other
animal species. Furthermore, member states are authorised to
use by way of derogation up to 31st December, 1980, complete
feedingstuffs for pigs with a maximum copper content of 200
mg/kg.

The Commission intends, however, to propose a postpone-
ment of this derogation up to 30th November, 1981. A decision
is thus to be taken in the near future as to the conditions of
use of copper salts in feedingstuff acceptable at Community
level. For this purpose, the opinion of the Scientific Committee for Animal Nutrition and Growth will be sought on the safety
of copper residues in animal tissues and organs for the consumer, and the consequences for the environment of the presence

of copper excreted by animals.

Since these questions were also the subject of the programme of the present Workshop, one may hope that complete
knowledge of the matter will be available in order to help the
Commission in its proposals."

FINAL DISCUSSION

Dr. Gasser presented draft Conclusions and Recommendations for comment and discussion. The drafts were circulated
to all participants immediately after the Workshop and correct-
ions, comments, deletions, additions and modifications were
invited. All comments made at the Workshop and by correspond-
ence have been incorporated in the appended Conclusions and
Recommendations.

CONCLUSIONS AND RECOMMENDATIONS

The Workshop examined the use of copper as a growth promoter for pigs and the consequences to the environment of the resulting slurry with enhanced contents of copper. Simultaneously consideration was given to the effects of sewage sludge containing copper on the soil, the plant and the animal.

The Workshop reached the following conclusions:

1)   The weight of evidence favours the continued use of copper as a feed additive in order to improve the efficiency of pig production. The information on the amounts needed and the ages of the pigs at which it should be administered is less clear cut, especially whether the maximum level required is 125 mg/kg or 200 mg/kg. Therefore there is insufficient evidence to justify altering the present position on scientific grounds.

2)   The problems of the use of copper as a growth promoter in pig feeds are long term in nature and will require considerable further investigation to solve them.

3)   Copper is required for the development of all animals but the mechanisms by which added copper promotes growth in pigs are not fully understood.

4)   Copper is known to interact strongly with other elements in animal nutrition, for example the administration of higher levels of zinc and iron decreases the toxic effects of copper and its level in the liver. With ruminants there are very marked interactions between copper, molybdenum and sulphur. Such interactions can influence the effects of copper applied by slurry or sludge to pasture on the health and performance of grazing animals.

5)   The forms of copper in animal faeces and slurry and sludge are not completely known nor their availability to plants or to animals when grazing contaminated swards.

6)   All copper applied to land may not always be accounted for by analysis of the soils. The errors in assessing

the amounts applied and in sampling and analysing the soil should be assessed. Any pathways of loss should be identified, including the forms of copper that may be lost by various mechanisms.

7) The effects of two or more heavy metals may be additive, independent or interactive. In the absence of detailed information on multi-element systems the use of the additive concept for applying copper and other heavy metals to soils tends to lead to more conservative applications.

8) For the consumer, the enhanced levels of copper in the livers of pigs receiving supplementary copper are no greater than in some other animal livers, and do not adversely affect the health of consumers.

9) If the present policy of disposing of animal wastes on limited areas is continued, the application of slurry from pigs receiving supplementary copper could increase the concentrations in the soil to unacceptably high levels within a finite time. The use of lower levels of supplementation will increase the time before toxic amounts are present but the major factor in determining the amount of copper applied to the land is the intensity of pig production in the area.

10) The most desirable solution to this problem is to remove the copper from the slurry before disposal to land if this becomes feasible and economic. No current work is known which would allow this, but a close watch should be kept on appropriate technology.

11) Although some experimental work has demonstrated that the ingestion by sheep of herbage heavily contaminated with slurry from pigs receiving added dietary copper can cause toxicity, these problems can be overcome by proper slurry management and animal husbandry. Breeds of sheep vary much in their susceptibility to deficiency and excess of copper in the diet, and lambs more so than sheep.

12) A uniform solution for all countries is difficult to foresee because of variations in conditions. For example, the large areas of copper deficient soils would be expected to benefit from additions of copper as salts, in slurry or in sewage sludge, whereas no further copper should be added in other areas.

Recommendations

1) There is no scientific basis to recommend changing the existing directives on the use of copper as a feed additive for pigs; Annex II of Directive 70/254 permits up to 200 mg/kg.

2) Until further information is available on the behaviour of copper and other heavy metals in soils on plants and on animals, the framing of uniform conditions for land disposal cannot be done.

3) Further investigations are urgently required into the mode of action of copper as a growth promoter and the behaviour of copper in slurry and sewage sludge when applied to land.

4) Investigations are needed on the forms of copper in pig slurry and sewage sludge with a view to devising methods for their removal.

5) The technological problems of removing copper from pig slurry should be put to those concerned with these areas of work.

6) There is a need to continue looking for alternative growth promoters, which will substitute directly for copper and have no environmental impact.

LIST OF PARTICIPANTS

Dr. M. Ahtiainen   North Karelia Water District Office
PO Box 69
80101 Joensvu 10
FINLAND

Dr. A. Aumaitre   Pig Husbandry Department,
Centre de Rennes-St-Gilles
35590 L'Hermitage
FRANCE

Dr. P.H.T. Beckett   Department of Agricultural Science
University of Oxford
Parks Road
Oxford
UK

Mr. A.D. Bird   Agricultural Trade Consultant
Woodbury
10 Bromley Lane
Chislehurst
Kent BR7 6LE
UK

Dr. G. Bories   INRA Laboratoire de Recherches sur les
Additifs Alimentaires
180 Chemin de Tournefeuille
31300 Toulouse
FRANCE

Dr. R. Braude   CAB Pig News and Information
Lane End House
Shinfield
Reading
UK

Dr. I. Bremner   Rowett Research Institute
Bucksburn
Aberdeen
UK

Dr. B.C. Cooke   Dalgety Spillers Ltd.
The Promenade
Clifton
Bristol
UK

Dr. M. Coppenet   INRA Station d'agronomie
4 rue de Stang Vihan
29000 Quimper
FRANCE

Prof. S. Coppola   University of Naples
                            Institute of Agricultural Microbiology
                            80055 Portici
                            ITALY

Dr. A. Dam Kofoed   State Experimental Station
                            Vejen
                            DENMARK

Dr. R.D. Davis   Water Research Centre
                            Elder Way
                            Stevenage SG1 3RA
                            Herts
                            UK

Mr. J. Dehandtschutter   Commission of the European Communities
                            DG VI
                            200 rue de la Loi
                            B-1049 Brussels
                            BELGIUM

Dr. J. Delas   INRA Station d'agronomie
                            Centre de Recherches de Bordeaux
                            33140 Pont de la Maye
                            FRANCE

Dr. G.A. Fleming   The Agricultural Institute
                            Johnstown Castle
                            Wexford
                            IRELAND

Dr. J.K.R. Gasser   Agricultural Research Council
                            160 Great Portland Street
                            London W1N 6DT
                            UK

Dr. A. Gomez   INRA
                            La Grande Ferrade
                            33140 Pont de la Maye
                            FRANCE

Dr. S. Gupta   Swiss Federal Research Station
                               for Agricultural Chemistry
                            3097 Liebefeld-Berne
                            SWITZERLAND

Dr. P. L'Hermite   Commission of the European Communities
                            DG XII
                            200 rue de la Loi
                            B-1049 Brussels
                            BELGIUM

Mr. T.W.G. Hucker   Department of the Environment
                            2 Marsham Street
                            London SW1P 3EB
                            UK

Dr. C. Juste	INRA Station d'agronomie Centre de Recherches de Bordeaux 33140 Pont de la Maye FRANCE
Dr. L. Kiekens	State University of Ghent Coupure Links 533 B-9000 Ghent BELGIUM
Dr. M. Lamand	Laboratoire des Maladies Nutritionelles INRA - CRZV de Theix 63310 Beaumont FRANCE
Dr. N.P. Lenis	Institute for Livestock Feeding and Nutrition Research (IVVO) PO Box 160 8200 AD Lelystad THE NETHERLANDS
Dr. R. Leschber	Institut für Wasser-, Boden- und Lufthygiene Corrensplatz 1 D-1000 Berlin 33 FEDERAL REPUBLIC OF GERMANY
Dr. Th. M. Lexmond	Vakgroep Bodemkunde en Bemestingsleer Landbouwhogeschool De Dreyen 3 6703 BC Wageningen THE NETHERLANDS
Mrs. M. Limères	INRA Station d'agronomie Centre de Recherches de Bordeaux 33140 Pont de la Maye FRANCE
Dr. O. Lindgren	Swedish Environment Protection Board Box 1302 17125 Solna SWEDEN
Dr. J.B. Ludvigsen	National Institute of Animal Science Rolighedsvej 25 DK-1958 Copenhagen V DENMARK
Dr. A. Madsen	National Institute of Animal Science Rolighedsvej 25 DK-1958 Copenhagen V DENMARK
Dr. A. Manrique	Instituto Nacional de Investigaciones Agrarias Aptdo 178 Burgos SPAIN

Dr. D. McGrath	The Agricultural Institute Johnstown Castle Wexford IRELAND
Dr. A. Minderhoud	National Institute of Public Health PO Box 1 Bilthoven THE NETHERLANDS
Dr. D.B.R. Poole	The Agricultural Institute Dunsinea Castleknock Co. Dublin IRELAND
Mrs. M.J. Robins	Janssen Services 33a High Street Chislehurst Kent BR7 5AE UK
Dr. K.L. Robinson	1 Bovingden Heights Spinfield Lane Marlow SL7 2JR Bucks UK (Member EEC Scientific Committee for Animal Nutrition)
Dr. S. Rodriguez	Instituto Nacional de Investigaciones Agrarias Aptdo 178 Burgos SPAIN
Miss. C.M. Swale	Janssen Services 33a High Street Chislehurst Kent BR7 5AE UK
Dr. R.J. Unwin	Ministry of Agriculture, Fisheries and Food Great Westminster House Horseferry Road London SW1 UK
Ir. J.H. Voorburg	Rijks Agrarische Afvalwaterdienst Kemperbergerweg 67 Arnhem THE NETHERLANDS
Dr. M.D. Webber	Wastewater Technology Center PO Box 5050 Burlington Ontario CANADA L7R 4A6

Dr. P. Worthington

Department of the Environment
Room C5/03
2 Marsham Street
London SW1P 3EB
UK